Contemporary's
NUMBER POWER 2

Fractions, Decimals and Percents

JERRY HOWETT

CONTEMPORARY
BOOKS

CHICAGO

Published by Contemporary Books, Inc.
Two Prudential Plaza, Chicago, Illinois 60601-6790
Manufactured in the United States of America
International Standard Book Number: 0-8092-8010-8

Published simultaneously in Canada by
Fitzhenry & Whiteside
195 Allstate Parkway
Markham, Ontario L3R 4T8
Canada

Contents

Building Number Power

Fraction Skills Inventory   2

What Are Fractions?   5

Writing Fractions   6

Forms of Fractions   7

Reducing Fractions   8

Raising Fractions to Higher Terms   10

Changing Improper Fractions to Whole or Mixed Numbers   11

Changing Mixed Numbers to Improper Fractions   12

Adding Fractions with the Same Bottom Numbers   13

Adding Fractions with Different Bottom Numbers   16

    Finding a Common Denominator   17

Addition of Fractions: Applying Your Skills   20

Subtracting Fractions with the Same Bottom Numbers   21

Subtracting Fractions with Different Bottom Numbers   22

Borrowing and Subtracting Fractions   24

Subtraction of Fractions: Applying Your Skills   27

Multiplying Fractions   28

Canceling and Multiplying Fractions   29

Multiplying Fractions by Whole Numbers   31

Multiplying with Mixed Numbers   32

Multiplication of Fractions: Applying Your Skills   33

Dividing Fractions by Fractions   34

Dividing Whole Numbers by Fractions   36

Dividing Fractions by Whole Numbers   38

Dividing with Mixed Numbers   40

Division of Fractions: Applying Your Skills   42

Final Fraction Skills Inventory   43

Decimal Skills Inventory   46

What Are Decimals?   49

Reading Decimals   50

Writing Decimals   51

Changing Decimals to Fractions   52

Changing Fractions to Decimals   53

CONTENTS

Comparing Decimals   54
Adding Decimals   55
Addition of Decimals: Applying Your Skills   56
Subtracting Decimals   57
Subtraction of Decimals: Applying Your Skills   58
Multiplying Decimals   59
Multiplying Decimals by 10, 100, and 1,000   62
Multiplication of Decimals: Applying Your Skills   63
Dividing Decimals by Whole Numbers   64
Dividing Decimals by Decimals   65
Dividing Whole Numbers by Decimals   67
Dividing Decimals by 10, 100, and 1,000   68
Division of Decimals: Applying Your Skills   69
Final Decimal Skills Inventory   70

Percent Skills Inventory   73
What Are Percents?   76
Changing Decimals to Percents   76
Changing Percents to Decimals   77
Changing Fractions to Percents   78
Changing Percents to Fractions   79
Common Percents and Their Values as Proper Fractions   80
Finding a Percent of a Number   81
Finding a Percent of a Number: Applying Your Skills   83
Finding What Percent One Number Is of Another   85
Finding What Percent One Number Is of Another: Applying Your Skills   87
Finding a Number When a Percent of It Is Given   89
Finding a Number When a Percent of It Is Given: Applying Your Skills   91
Final Percent Skills Inventory   93

Building Number Power: Review Test   96

Using Number Power

Reading a Ruler   102
Reading a Metric Ruler   103
Perimeter: Finding the Distance Around Something (Rectangles and Squares)   104
Area: Finding the Amount of Space Taken Up by Something
    (Rectangles and Squares)   106
Volume: Finding the Amount of Space Inside Something (Rectangular Shapes)   108
Circumference: Finding the Distance Around a Circle   110
Area: Finding the Amount of Space Taken Up by a Circle   112

iv

Changing a Recipe   114

How Much Do you Pay? (Rounding Off Money to the Nearest Cent)   115

Finding Interest for One Year   118

Finding Interest for Less Than One Year   120

Finding Interest for More Than One Year   121

Finding Compound Interest   122

Comparing Food Prices: Unit Pricing   124

Finding the Percent Saved at a Sale   126

Buying Furniture on Sale   128

Using a Tax Rate Schedule   130

Filling Out a Wage and Tax Statement   132

Working with a Budget   134

Buying on the Installment Plan   136

To the Learner

Number Power 2 is the second workbook in the Number Power series.

The first section of the workbook, BUILDING NUMBER POWER, provides easy-to-follow instructions and plenty of practice in working with fractions, decimals, and percents. Each new section begins with a skills inventory so that you can evaluate your strengths and weaknesses and ends with a final skills inventory so that you can measure your progress.

The second part of the workbook, USING NUMBER POWER, gives you an opportunity to use your math skills to solve real-life math problems.

For both sections, the answers to the problems appear in the back of the book.

After carefully working your way through this workbook, you will be able to handle the fractions, decimals, and percents needed in daily living, in school, and on many tests including the GED. The math skills that you develop will give you a solid foundation for a future with number power.

Building
Number Power

Fraction Skills Inventory

Do all of the following problems that you can. There is no time limit. Work accurately, but do not use outside help. All answers should be written in lowest terms.

1. A pound contains 16 ounces. 10 ounces is what fraction of a pound?

2. Reduce $\frac{28}{42}$

3. Change $\frac{24}{10}$ to a mixed number.

4. Change $5\frac{3}{7}$ to an improper fraction.

5. $\begin{aligned} & 3\frac{11}{15} \\ +\, & 4\frac{7}{15} \end{aligned}$

6. $\begin{aligned} & 3\frac{1}{3} \\ & 2\frac{3}{4} \\ +\, & 5\frac{5}{6} \end{aligned}$

7. A wooden crate weighing $2\frac{5}{16}$ pounds contains grapefruit weighing $24\frac{1}{2}$ pounds. What is the combined weight of the crate and the grapefruit?

8. Fred, a traveling salesman, spent $1\frac{3}{4}$ hours driving on Monday, $2\frac{1}{3}$ hours driving on Tuesday, and $2\frac{1}{6}$ hours driving on Wednesday. What total number of hours did he spend driving those three days?

9. $12\frac{3}{4}$
   $- \ 6\frac{2}{5}$

10. $9\frac{1}{8}$
    $-4\frac{5}{8}$

11. $10\frac{2}{7}$
    $- \ 6\frac{2}{3}$

12. Maxine weighed 135 pounds. By dieting she lost $10\frac{1}{2}$ pounds. How much did she weigh after the diet?

13. From a piece of cloth $42\frac{1}{4}$ inches long, Celeste cut a strip $14\frac{3}{4}$ inches long. How long was the remaining piece?

14. $\frac{7}{10} \times \frac{3}{4} =$

15. $\frac{3}{5} \times \frac{10}{21} \times \frac{7}{8} =$

16. $2\frac{2}{5} \times 6\frac{1}{4} =$

17. Bill shipped $\frac{7}{10}$ of the crates in his warehouse by air freight. If there were 40 crates in the warehouse, how many did he ship?

18. If Gloria works $7\frac{1}{2}$ hours a day, how many hours does she work in a 5-day work week?

19. $\frac{5}{8} \div \frac{9}{16} =$

20. $12 \div \frac{3}{4} =$

3

21. $\frac{7}{9} \div 14 =$

22. $3\frac{3}{4} \div 1\frac{5}{7} =$

23. How many $\frac{1}{2}$ pound bags can be filled with 14 pounds of peanuts?

24. From a strip of wood 125 inches long, George is cutting pieces $12\frac{1}{2}$ inches long. How many pieces can he cut from the long strip?

FRACTION SKILLS INVENTORY CHART

A passing score is 20 problems correct. If you had less than 20 correct, complete pages 5 through 45 before going on. Even if you had a passing score or better, any problem you missed should be corrected. Following is a list of the problems and the pages where each problem is covered.

Problem Number	Practice Page	Problem Number	Practice Page
1	6	13	27
2	8–9	14	28
3	11	15	29–30
4	12	16	32
5	13–15	17	33
6	16–19	18	33
7	20	19	34–35
8	20	20	36–37
9	22–23	21	38–39
10	24–26	22	40–41
11	24–26	23	42
12	27	24	42

What Are Fractions?

A **fraction** is a part of something. A penny is a fraction of a dollar. It is one of the 100 equal parts of a dollar or $\frac{1}{100}$ (one hundred*th*) of a dollar. An inch is a fraction of a foot. It is one of the 12 equal parts of a foot or $\frac{1}{12}$ (one twelf*th*) of a foot. 5 days are a fraction of a week. They are 5 of the 7 equal parts of a week or $\frac{5}{7}$ (five seven*ths*) of a week.

The two numbers in a fraction are called the

$$\frac{\text{numerator}}{\text{denominator}}$$ —which tells how many parts you have
　　　　　　　　　　—which tells how many parts in the whole

EXAMPLE: The fraction $\frac{3}{4}$ tells you what part of the figure at the right is shaded. 3 parts are shaded. The whole figure is divided into 4 equal parts.

Write fractions that represent the part of each figure that is shaded.

1. _____

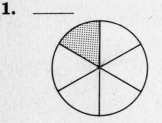

2. _____

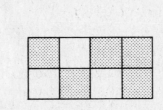

3. _____

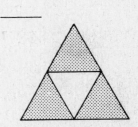

4. _____

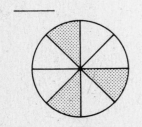

5. _____

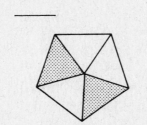

6. _____

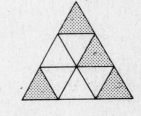

7. _____

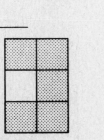

8. _____

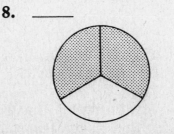

9. _____

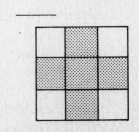

Writing Fractions

Since there are three feet in one yard, one foot is $\frac{1}{3}$ of a yard. Two feet make up $\frac{2}{3}$ of a yard.

Write fractions for each of the *parts* described below.

_____ 1. A foot contains 12 inches. 5 inches is what fraction of a foot?

_____ 2. 47¢ is what fraction of a dollar?

_____ 3. A pound contains 16 ounces. 9 ounces is what fraction of a pound?

_____ 4. A yard contains 36 inches. 23 inches is what fractional part of a yard?

_____ 5. 7 months is what fraction of a year?

_____ 6. 8¢ is what fraction of a quarter?

_____ 7. Earlene wants a coat that costs $60. She has saved $43. What fraction of the amount that she needs has she saved?

_____ 8. There are 2,000 pounds in a ton. 1,351 pounds is what fraction of a ton?

_____ 9. During a 5-day work week, Pete was sick for 2 days. What fraction of the work week was he sick?

_____10. There are 100 centimeters in a meter. 63 centimeters is what fraction of a meter?

_____11. There are 4 quarts in a gallon. 3 quarts is what fraction of a gallon?

_____12. David makes $150 a week. He has spent $113. What fraction of his week's pay has he spent?

_____13. Madge has typed 77 pages of a report that contains 280 pages. What fraction of the report has she typed?

Forms of Fractions

Proper fraction —The top number is *less than* the bottom number.

EXAMPLES: $\frac{1}{3}, \frac{3}{10}, \frac{7}{19}$

A proper fraction is less than all the parts the whole is divided into. The value of a proper fraction is *always less than one.*

Improper fraction —The top number is *equal to or larger than* the bottom number.

EXAMPLES: $\frac{3}{2}, \frac{9}{4}, \frac{8}{8}$

An improper fraction is all the parts that a whole is divided into such as $\frac{8}{8}$, or it is more than the total parts in the whole. The value of an improper fraction is either *equal to one or more than one.*

Mixed number —A whole number is written next to a proper fraction.

EXAMPLES: $1\frac{2}{5}, 3\frac{1}{2}, 10\frac{4}{7}$

Tell whether each of the following is a proper fraction (P), an improper fraction (I), or a mixed number (M).

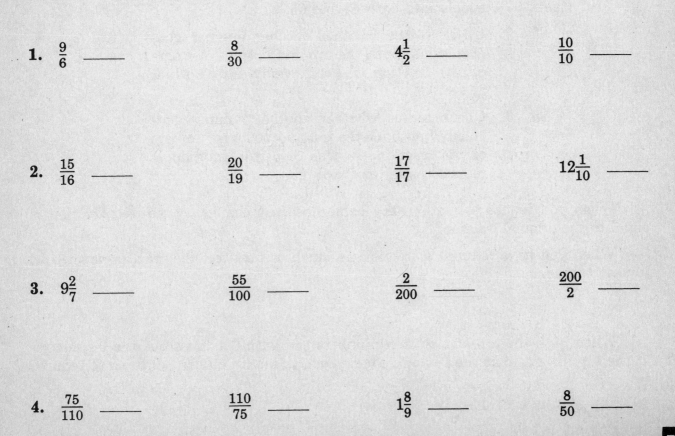

1. $\frac{9}{6}$ _____ $\frac{8}{30}$ _____ $4\frac{1}{2}$ _____ $\frac{10}{10}$ _____

2. $\frac{15}{16}$ _____ $\frac{20}{19}$ _____ $\frac{17}{17}$ _____ $12\frac{1}{10}$ _____

3. $9\frac{2}{7}$ _____ $\frac{55}{100}$ _____ $\frac{2}{200}$ _____ $\frac{200}{2}$ _____

4. $\frac{75}{110}$ _____ $\frac{110}{75}$ _____ $1\frac{8}{9}$ _____ $\frac{8}{50}$ _____

Reducing Fractions

The coin we call a quarter stands for 25 pennies out of the total of 100 pennies in a dollar. You could call 25¢ $\frac{25}{100}$ of a dollar, or thinking of the 5 nickels in 25¢, $\frac{5}{20}$ of a dollar. The easiest way is to say that 25¢ is $\frac{1}{4}$ of a dollar. $\frac{1}{4}$ is the *reduced* form of $\frac{25}{100}$ and $\frac{5}{20}$. Reducing a fraction means writing it an easier way—with smaller numbers.

Study the following examples to see how fractions are reduced.

EXAMPLE 1.   Reduce $\frac{15}{20}$

Step 1. Find a number that goes evenly into the top and bottom numbers of the fraction. 5 goes evenly into both 15 and 20.

$$\frac{15 \div 5}{20 \div 5} = \frac{3}{4}$$

Step 2. Check to see whether another number goes evenly into both the top and bottom numbers of the fraction. Since no other number goes evenly into both 3 and 4, the fraction is reduced as far as it will go.

EXAMPLE 2.   Reduce $\frac{48}{64}$

Step 1. Find a number that goes evenly into the top and bottom numbers of the fraction. 8 goes evenly into both 48 and 64.

$$\frac{48 \div 8}{64 \div 8} = \frac{6}{8}$$

Step 2. Check to see whether another number goes evenly into both the top and bottom numbers of the fraction. 2 goes evenly into both 6 and 8.

$$\frac{6 \div 2}{8 \div 2} = \frac{3}{4}$$

Step 3. Check to see whether another number goes evenly into both the top and bottom numbers of the fraction. In this case, the fraction is reduced as far as it will go.

When you reduce a fraction, the value does not change. A reduced fraction is equal to the original fraction.

When you have reduced a fraction as much as possible, the fraction is then in **lowest terms.**

When both the top and bottom numbers end with 0's, cross out the 0's, a zero at the top for a zero at the bottom. Then check to see if you can continue to reduce.

EXAMPLE 3.   Reduce $\frac{20}{30}$

Step 1. Cross out the 0's at the end of each number. $\dfrac{2\cancel{0}}{3\cancel{0}}$

Step 2. Check to see if you can continue to reduce. In this case, the fraction is reduced as far as it will go. $\dfrac{2}{3}$

EXAMPLE 4. Reduce $\dfrac{40}{100}$

Step 1. Cross out one 0 at the end of each number. Be sure that you cross out only one zero in the bottom number since you crossed out only one zero in the top number. $\dfrac{4\cancel{0}}{10\cancel{0}}$

Step 2. Check to see if you can continue to reduce. 2 goes evenly into both 4 and 10. $\dfrac{4 \div 2}{10 \div 2} = \dfrac{2}{5}$

Reduce each fraction to lowest terms.

1. $\dfrac{6}{12} =$ $\dfrac{7}{28} =$ $\dfrac{3}{9} =$ $\dfrac{9}{45} =$ $\dfrac{6}{48} =$

2. $\dfrac{25}{30} =$ $\dfrac{32}{36} =$ $\dfrac{16}{18} =$ $\dfrac{21}{24} =$ $\dfrac{14}{21} =$

3. $\dfrac{20}{50} =$ $\dfrac{30}{90} =$ $\dfrac{70}{200} =$ $\dfrac{90}{140} =$ $\dfrac{80}{170} =$

4. $\dfrac{33}{77} =$ $\dfrac{45}{60} =$ $\dfrac{18}{36} =$ $\dfrac{42}{56} =$ $\dfrac{48}{64} =$

5. $\dfrac{75}{80} =$ $\dfrac{420}{480} =$ $\dfrac{72}{90} =$ $\dfrac{26}{39} =$ $\dfrac{18}{32} =$

6. $\dfrac{25}{50} =$ $\dfrac{14}{42} =$ $\dfrac{4}{200} =$ $\dfrac{63}{81} =$ $\dfrac{35}{49} =$

Raising Fractions to Higher Terms

An important skill in addition and subtraction of fractions is raising a fraction to **higher terms.** This is the opposite of reducing a fraction to **lowest terms.**

EXAMPLE 1. Raise $\frac{2}{5}$ to 20ths.

Step 1. Divide the old bottom number into the new one.

$$5\overline{)20}^{\,4}$$

Step 2. Multiply the answer (4) by the old top number (2).

$$\frac{2 \times 4}{5 \times 4} = \frac{8}{20}$$

Check. Reduce the new fraction to see if you get the original fraction.

$$\frac{8 \div 4}{20 \div 4} = \frac{2}{5}$$

EXAMPLE 2. $\frac{4}{9} = \frac{?}{27}$

Step 1. Divide the old bottom number into the new one.

$$9\overline{)27}^{\,3}$$

Step 2. Multiply the answer (3) by the old top number (4).

$$\frac{4 \times 3}{9 \times 3} = \frac{12}{27}$$

Check. Reduce the new fraction to see if you get the original fraction.

$$\frac{12 \div 3}{27 \div 3} = \frac{4}{9}$$

Raise each fraction to higher terms by filling in the missing top number.

1. $\frac{4}{5} = \frac{}{30}$ $\frac{9}{10} = \frac{}{20}$ $\frac{1}{6} = \frac{}{18}$ $\frac{5}{8} = \frac{}{32}$

2. $\frac{4}{7} = \frac{}{35}$ $\frac{1}{2} = \frac{}{36}$ $\frac{2}{3} = \frac{}{21}$ $\frac{9}{11} = \frac{}{66}$

3. $\frac{5}{9} = \frac{}{45}$ $\frac{3}{4} = \frac{}{44}$ $\frac{7}{12} = \frac{}{60}$ $\frac{1}{3} = \frac{}{45}$

Changing Improper Fractions to Whole or Mixed Numbers

An improper fraction is a fraction with a top number that is as big or bigger than the bottom number. An improper fraction is equal to or larger than one whole.

Suppose a group of people order two pizzas each of which has been cut into seven equal parts (see the illustration at the right). Each slice is then $\frac{1}{7}$ of the whole pizza. If two people eat seven slices between them (or $\frac{7}{7}$), they have eaten one whole pizza ($\frac{7}{7} = 1$). If those two people had eaten eight slices between them (or $\frac{8}{7}$), they would have eaten one whole pizza plus $\frac{1}{7}$ of another pizza ($\frac{8}{7} = 1\frac{1}{7}$).

You can change any improper fraction, such as $\frac{8}{7}$, into a mixed number by dividing the bottom number into the top number and writing the remainder, if any, over the original bottom number.

EXAMPLE: Change $\frac{21}{9}$ to a mixed number.

Step 1.   Divide the bottom into the top.

$$\begin{array}{r} 2 \\ 9{\overline{)}21} \\ 18 \\ \hline 3 \end{array}$$

Step 2.   Write the remainder as a fraction over the original bottom number.

$2\frac{3}{9}$

Step 3.   Reduce the remaining fraction.

$\frac{3 \div 3}{9 \div 3} = \frac{1}{3}$

The answer becomes $2\frac{1}{3}$.

Change each fraction to a whole or mixed number. Be sure to reduce any remaining fractions.

1. $\frac{14}{8} =$ $\frac{33}{6} =$ $\frac{14}{5} =$ $\frac{30}{7} =$ $\frac{12}{3} =$

2. $\frac{30}{9} =$ $\frac{26}{8} =$ $\frac{18}{6} =$ $\frac{36}{10} =$ $\frac{16}{8} =$

3. $\frac{13}{12} =$ $\frac{45}{9} =$ $\frac{45}{6} =$ $\frac{32}{12} =$ $\frac{42}{9} =$

Changing Mixed Numbers to Improper Fractions

An important skill in multiplication and division of fractions is changing a mixed number, such as $2\frac{1}{4}$, to an improper fraction. One whole is equal to $\frac{4}{4}$. 2 is equal to $\frac{8}{4}$. Adding the extra $\frac{1}{4}$, we get $\frac{9}{4}$.

Study the following examples to see how mixed numbers are changed to improper fractions.

EXAMPLE 1.   Change $2\frac{1}{4}$ to an improper fraction.

Step 1.   Multiply the bottom number by the whole number. $4 \times 2 = 8$

Step 2.   Add the result to the top number. $8 + 1 = 9$

Step 3.   Place the total over the bottom number. $\frac{9}{4}$

EXAMPLE 2.   Change $5\frac{2}{3}$ to an improper fraction.

Step 1. $3 \times 5 = 15$

Step 2. $15 + 2 = 17$

Step 3.   Place 17 over 3

Answer: $5\frac{2}{3} = \frac{17}{3}$

Change each mixed number to an improper fraction.

1. $2\frac{3}{4} =$ $1\frac{4}{7} =$ $5\frac{1}{3} =$ $6\frac{2}{7} =$ $4\frac{3}{5} =$

2. $9\frac{1}{2} =$ $7\frac{5}{8} =$ $2\frac{9}{10} =$ $8\frac{3}{4} =$ $3\frac{5}{9} =$

3. $10\frac{1}{3} =$ $11\frac{2}{5} =$ $4\frac{5}{12} =$ $6\frac{7}{8} =$ $12\frac{1}{4} =$

Adding Fractions with the Same Bottom Numbers

To add fractions with the same bottom numbers, add the top numbers, and put the total over the bottom number.

EXAMPLE:

$$\frac{2}{7}$$

Step 1. Add the top numbers. $2 + 3 = 5$.

$$+\frac{3}{7}$$

Step 2. Place the total (5) over the bottom number: $\frac{5}{7}$

$$\frac{5}{7}$$

Add the following.

1. $\frac{2}{9}$ $+\frac{3}{9}$ $\frac{3}{7}$ $+\frac{1}{7}$ $\frac{4}{8}$ $+\frac{3}{8}$ $\frac{5}{12}$ $+\frac{2}{12}$ $\frac{4}{13}$ $+\frac{6}{13}$

2. $\frac{3}{11}$ $\frac{1}{11}$ $+\frac{2}{11}$ $\frac{5}{9}$ $\frac{2}{9}$ $+\frac{1}{9}$ $\frac{2}{15}$ $\frac{7}{15}$ $+\frac{4}{15}$ $\frac{8}{17}$ $\frac{2}{17}$ $+\frac{5}{17}$ $\frac{2}{19}$ $\frac{9}{19}$ $+\frac{5}{19}$

3. $4\frac{2}{5}$ $+3\frac{1}{5}$ $6\frac{3}{10}$ $+8\frac{6}{10}$ $5\frac{4}{11}$ $+4\frac{5}{11}$ $8\frac{7}{13}$ $+6\frac{4}{13}$

4. $3\frac{2}{9}$ $5\frac{1}{9}$ $+4\frac{5}{9}$ $6\frac{4}{11}$ $9\frac{2}{11}$ $+2\frac{3}{11}$ $7\frac{2}{7}$ $8\frac{2}{7}$ $+5\frac{2}{7}$ $9\frac{3}{10}$ $2\frac{5}{10}$ $+4\frac{1}{10}$

Sometimes the total of an addition problem can be reduced.

EXAMPLE:

$$\frac{5}{12}$$

Step 1. Add the top numbers. $5 + 1 = 6$.

$$+\frac{1}{12}$$

Step 2. Place the total over the bottom number: $\frac{6}{12}$

$$\frac{6}{12} = \frac{1}{2}$$

Step 3. Reduce the answer: $\frac{6 \div 6}{12 \div 6} = \frac{1}{2}$

Add and reduce.

5.
$$\frac{5}{8}$$
$$+\frac{1}{8}$$

$$\frac{4}{9}$$
$$+\frac{2}{9}$$

$$\frac{3}{12}$$
$$+\frac{5}{12}$$

$$\frac{7}{15}$$
$$+\frac{3}{15}$$

$$\frac{3}{16}$$
$$+\frac{5}{16}$$

6.
$$\frac{2}{10}$$
$$\frac{3}{10}$$
$$+\frac{3}{10}$$

$$\frac{6}{15}$$
$$\frac{2}{15}$$
$$+\frac{4}{15}$$

$$\frac{7}{20}$$
$$\frac{3}{20}$$
$$+\frac{4}{20}$$

$$\frac{9}{24}$$
$$\frac{7}{24}$$
$$+\frac{2}{24}$$

$$\frac{8}{27}$$
$$\frac{4}{27}$$
$$+\frac{6}{27}$$

7.
$$5\frac{2}{6}$$
$$+6\frac{1}{6}$$

$$7\frac{5}{12}$$
$$+9\frac{3}{12}$$

$$10\frac{5}{14}$$
$$+\ 8\frac{7}{14}$$

$$13\frac{8}{15}$$
$$+29\frac{2}{15}$$

8.
$$9\frac{2}{9}$$
$$8\frac{1}{9}$$
$$+7\frac{3}{9}$$

$$6\frac{3}{10}$$
$$7\frac{2}{10}$$
$$+4\frac{3}{10}$$

$$10\frac{5}{16}$$
$$4\frac{3}{16}$$
$$+3\frac{4}{16}$$

$$12\frac{3}{8}$$
$$9\frac{1}{8}$$
$$+10\frac{2}{8}$$

If the total of an addition problem is an improper fraction, change the total to a mixed number. (See page 11.)

EXAMPLE:

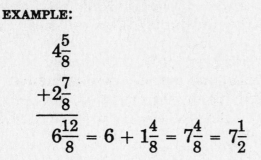

$$4\frac{5}{8}$$
$$+2\frac{7}{8}$$
$$6\frac{12}{8} = 6 + 1\frac{4}{8} = 7\frac{4}{8} = 7\frac{1}{2}$$

Step 1. Add the top numbers.
$5 + 7 = 12$.

Step 2. Place the total over the bottom number: $\frac{12}{8}$

Step 3. Change the improper fraction to a mixed number: $\frac{12}{8} = 1\frac{4}{8}$, and add the whole number to the whole number column.
$6 + 1\frac{4}{8} = 7\frac{4}{8}$.

Step 4. Reduce the remainder: $7\frac{4}{8} = 7\frac{1}{2}$.

Add and reduce.

9.
$\frac{4}{5}$
$+\frac{3}{5}$

$\frac{6}{8}$
$+\frac{5}{8}$

$\frac{7}{10}$
$+\frac{6}{10}$

$\frac{8}{9}$
$+\frac{5}{9}$

$\frac{3}{6}$
$+\frac{3}{6}$

10.
$\frac{11}{12}$
$+\frac{5}{12}$

$\frac{9}{14}$
$+\frac{7}{14}$

$\frac{7}{8}$
$+\frac{5}{8}$

$\frac{11}{15}$
$+\frac{7}{15}$

$\frac{9}{16}$
$+\frac{15}{16}$

11.
$\frac{5}{9}$
$\frac{7}{9}$
$+\frac{8}{9}$

$\frac{6}{7}$
$\frac{5}{7}$
$+\frac{3}{7}$

$\frac{7}{10}$
$\frac{3}{10}$
$+\frac{5}{10}$

$\frac{9}{12}$
$\frac{11}{12}$
$+\frac{7}{12}$

$\frac{6}{8}$
$\frac{5}{8}$
$+\frac{7}{8}$

12.
$8\frac{3}{8}$
$3\frac{7}{8}$
$+5\frac{5}{8}$

$9\frac{5}{6}$
$2\frac{1}{6}$
$+6\frac{4}{6}$

$5\frac{7}{10}$
$4\frac{9}{10}$
$+7\frac{5}{10}$

$2\frac{7}{12}$
$8\frac{8}{12}$
$+6\frac{9}{12}$

Adding Fractions with Different Bottom Numbers

If the fractions in an addition problem do not have the same bottom numbers (**denominators**), you must rewrite the problem so that all of the fractions have the same bottom number (called a **common denominator**). This will mean raising at least one of the fractions to higher terms (see page 10).

A common denominator is a number that can be divided evenly by all of the denominators in the problem. The smallest number that can be divided evenly by all of the denominators in the problem is called the **lowest common denominator** or **LCD**. Sometimes, the largest denominator in the problem will work as the LCD.

EXAMPLE 1

$$\frac{3}{5} = \frac{9}{15}$$

Step 1. Since 5 divides evenly into 15, 15 is the LCD.

$$+\frac{4}{15} = \frac{4}{15}$$

Step 2. Raise $\frac{3}{5}$ to 15ths.

$$\frac{13}{15}$$

Step 3. Add the new fractions.

Add and reduce.

1.
$$\frac{3}{4} \quad\quad \frac{2}{3} \quad\quad \frac{7}{8} \quad\quad \frac{5}{6} \quad\quad \frac{5}{9}$$
$$+\frac{1}{2} \quad\quad +\frac{5}{6} \quad\quad +\frac{3}{4} \quad\quad +\frac{1}{3} \quad\quad +\frac{2}{3}$$

2.
$$\frac{3}{8} \quad\quad \frac{1}{6} \quad\quad \frac{2}{5} \quad\quad \frac{2}{3} \quad\quad \frac{3}{5}$$
$$\frac{3}{4} \quad\quad \frac{5}{12} \quad\quad \frac{1}{2} \quad\quad \frac{5}{12} \quad\quad \frac{1}{3}$$
$$+\frac{1}{2} \quad\quad +\frac{3}{4} \quad\quad +\frac{9}{10} \quad\quad +\frac{1}{4} \quad\quad +\frac{4}{15}$$

3.
$$\frac{3}{4} \quad\quad \frac{2}{3} \quad\quad \frac{5}{24} \quad\quad \frac{4}{9} \quad\quad \frac{7}{10}$$
$$\frac{1}{2} \quad\quad \frac{5}{6} \quad\quad \frac{3}{8} \quad\quad \frac{5}{6} \quad\quad \frac{1}{3}$$
$$+\frac{3}{20} \quad\quad +\frac{1}{2} \quad\quad +\frac{1}{3} \quad\quad +\frac{7}{18} \quad\quad +\frac{11}{30}$$

4.

$$14\tfrac{4}{5}$$
$$+7\tfrac{8}{15}$$

$$7\tfrac{5}{9}$$
$$+6\tfrac{11}{18}$$

$$9\tfrac{1}{2}$$
$$+5\tfrac{15}{24}$$

$$3\tfrac{3}{7}$$
$$+18\tfrac{31}{42}$$

5.

$$7\tfrac{2}{3}$$
$$3\tfrac{1}{2}$$
$$+9\tfrac{5}{6}$$

$$8\tfrac{5}{8}$$
$$5\tfrac{1}{4}$$
$$+4\tfrac{3}{16}$$

$$6\tfrac{1}{2}$$
$$2\tfrac{9}{10}$$
$$+7\tfrac{4}{5}$$

$$4\tfrac{3}{4}$$
$$8\tfrac{5}{12}$$
$$+3\tfrac{5}{6}$$

FINDING A COMMON DENOMINATOR

Here are some ways of finding a common denominator when the largest denominator in an addition problem doesn't work:

A. Multiply the denominators together.

B. Go through the multiplication table of the largest denominator.

EXAMPLE 2

$$\frac{2}{5} = \frac{8}{20}$$
$$+\frac{3}{4} = \frac{15}{20}$$
$$\overline{\frac{23}{20}} = 1\tfrac{3}{20}$$

Step 1. Multiply the denominators.
$5 \times 4 = 20.$ 20 is the LCD.

Step 2. Raise each fraction to 20ths as on page 10.

Step 3. Add as usual (see pages 13, 14, and 15).

Step 4. Change the answer to a mixed number.

EXAMPLE 3

$$\frac{2}{3} = \frac{8}{12}$$
$$\frac{5}{6} = \frac{10}{12}$$
$$+\frac{3}{4} = \frac{9}{12}$$
$$\overline{\frac{27}{12}} = 2\tfrac{3}{12} = 2\tfrac{1}{4}$$

Step 1. Go through the multiplication table of the 6's.
$6 \times 1 = 6$, which cannot be divided by 4.
$6 \times 2 = 12$, which can be divided by 3 and 4.

Step 2. Raise each fraction to 12ths.

Step 3. Add.

Step 4. Change the answer to a mixed number and reduce.

ADDING FRACTIONS WITH DIFFERENT BOTTOM NUMBERS

6. $\frac{3}{5}$ $+\frac{2}{3}$ $\frac{3}{4}$ $+\frac{1}{3}$ $\frac{2}{5}$ $+\frac{1}{2}$ $\frac{3}{7}$ $+\frac{1}{3}$ $\frac{5}{6}$ $+\frac{2}{5}$

7. $\frac{4}{7}$ $+\frac{3}{4}$ $\frac{5}{6}$ $+\frac{2}{7}$ $\frac{3}{8}$ $+\frac{4}{5}$ $\frac{2}{3}$ $+\frac{4}{11}$ $\frac{5}{9}$ $+\frac{3}{5}$

8. $\frac{5}{6}$ $+\frac{3}{4}$ $\frac{4}{9}$ $+\frac{5}{6}$ $\frac{7}{10}$ $+\frac{3}{4}$ $\frac{5}{12}$ $+\frac{5}{9}$ $\frac{3}{8}$ $+\frac{5}{6}$

9. $\frac{2}{3}$ $\frac{5}{8}$ $+\frac{3}{4}$ $\frac{1}{4}$ $\frac{3}{5}$ $+\frac{7}{10}$ $\frac{8}{9}$ $\frac{5}{6}$ $+\frac{3}{4}$ $\frac{5}{16}$ $\frac{5}{8}$ $+\frac{1}{2}$ $\frac{2}{9}$ $\frac{1}{2}$ $+\frac{5}{6}$

10. $\frac{5}{6}$ $\frac{2}{5}$ $+\frac{4}{15}$ $\frac{7}{12}$ $\frac{5}{8}$ $+\frac{3}{4}$ $\frac{2}{3}$ $\frac{1}{6}$ $+\frac{11}{12}$ $\frac{7}{20}$ $\frac{3}{8}$ $+\frac{3}{10}$ $\frac{4}{9}$ $\frac{1}{6}$ $+\frac{5}{12}$

11. $\frac{2}{3}$ $\frac{4}{9}$ $+\frac{5}{6}$ $\frac{6}{7}$ $\frac{3}{4}$ $+\frac{1}{2}$ $\frac{2}{9}$ $\frac{1}{3}$ $+\frac{1}{6}$ $\frac{5}{7}$ $\frac{4}{9}$ $+\frac{2}{3}$ $\frac{11}{16}$ $\frac{1}{3}$ $+\frac{7}{8}$

12. $4\frac{3}{5}$ $7\frac{5}{8}$ $8\frac{5}{9}$ $6\frac{5}{12}$
   $+6\frac{3}{4}$ $+9\frac{2}{3}$ $+3\frac{2}{3}$ $+7\frac{3}{8}$

13. $10\frac{2}{7}$ $6\frac{1}{6}$ $7\frac{9}{10}$ $3\frac{4}{9}$
   $+\ 8\frac{1}{3}$ $+4\frac{3}{4}$ $+8\ \frac{1}{4}$ $+12\frac{5}{6}$

14. $9\ \frac{5}{8}$ $8\frac{5}{6}$ $3\frac{7}{8}$ $6\frac{9}{10}$
   $+3\frac{7}{12}$ $+2\frac{2}{9}$ $+11\frac{3}{5}$ $+5\ \frac{2}{3}$

15. $5\ \frac{2}{3}$ $3\frac{3}{8}$ $2\frac{2}{9}$ $6\frac{1}{2}$
   $9\ \frac{3}{5}$ $8\frac{1}{6}$ $10\frac{3}{4}$ $1\frac{1}{3}$
   $+2\frac{7}{10}$ $+7\frac{3}{4}$ $+\ 4\frac{5}{6}$ $+14\frac{1}{4}$

16. $7\frac{7}{8}$ $9\frac{5}{12}$ $6\frac{7}{16}$ $12\frac{3}{4}$
   $\frac{2}{3}$ $11\ \frac{2}{3}$ $\frac{1}{2}$ $9\frac{5}{6}$
   $+4\frac{1}{2}$ $+\ \ \frac{4}{9}$ $+8\ \frac{5}{8}$ $+\ \ \frac{2}{3}$

17. $9\frac{1}{2}$ $\frac{6}{7}$ $4\frac{5}{12}$ $11\frac{7}{8}$
   $3\frac{3}{4}$ $15\ \frac{3}{5}$ $\frac{4}{9}$ $5\frac{3}{7}$
   $+2\frac{2}{3}$ $+\ 7\frac{7}{10}$ $+19\ \frac{3}{4}$ $+\ \ \frac{1}{2}$

Addition of Fractions: Applying Your Skills

Addition problems generally ask you to **combine** figures or to find a **total** or **sum**. Be sure to reduce your answers to lowest terms.

1. Karen is $62\frac{1}{2}$ inches tall. Her mother is $5\frac{3}{4}$ inches taller. How tall is Karen's mother?

2. Doing errands on Monday, Mrs. Johnson drove $5\frac{1}{2}$ miles to the supermarket, $3\frac{7}{10}$ miles to the hardware store, $\frac{8}{10}$ mile to the laundry, and $6\frac{1}{10}$ miles back home. How far did she drive all together?

3. Mr. Munro's empty suitcase weighs $4\frac{3}{4}$ pounds. If the things that he puts in the suitcase weigh $17\frac{3}{5}$ pounds, what is the weight of the suitcase when it is filled?

4. When Petra went shopping she bought 2 pounds of sugar, $3\frac{1}{4}$ pounds of ground beef, $2\frac{2}{3}$ pounds of cheese, and a $\frac{7}{8}$-pound can of tomatoes. What was the total weight of her purchase?

5. John spends his evenings trying to change his attic into an extra bedroom. Monday night he worked $3\frac{1}{2}$ hours; Tuesday night he worked $4\frac{1}{3}$ hours; Wednesday he worked $2\frac{3}{4}$ hours; and Thursday he worked $3\frac{2}{3}$ hours. How many hours did he work on his attic that week?

6. When Ruby was sick, her weight went down to $116\frac{1}{5}$ pounds. By the time she recovered, she had gained $12\frac{1}{2}$ pounds. What was her final weight when she was well?

7. Lois talked on the phone for $\frac{1}{2}$ hour in the morning, $\frac{3}{5}$ of an hour in the afternoon, and $1\frac{2}{3}$ hours in the evening. How much time did she spend on the phone that day?

Subtracting Fractions with the Same Bottom Numbers

To subtract fractions with the same bottom numbers, subtract the top numbers and put the total over the bottom number.

EXAMPLE

$$\frac{5}{12}$$

$$-\frac{1}{12}$$

$$\frac{4}{12} = \frac{1}{3}$$

Step 1. Subtract the top numbers. $5 - 1 = 4$.

Step 2. Place the answer over the bottom number: $\frac{4}{12}$

Step 3. Reduce the answer: $\frac{4 \div 4}{12 \div 4} = \frac{1}{3}$

Subtract and reduce.

1.
$\frac{5}{9}$ $-\frac{2}{9}$ $\frac{7}{10}$ $-\frac{6}{10}$ $\frac{5}{8}$ $-\frac{1}{8}$ $\frac{4}{13}$ $-\frac{1}{13}$ $\frac{9}{11}$ $-\frac{3}{11}$

2.
$\frac{13}{15}$ $-\frac{8}{15}$ $\frac{15}{16}$ $-\frac{9}{16}$ $\frac{23}{24}$ $-\frac{11}{24}$ $\frac{11}{19}$ $-\frac{8}{19}$ $\frac{17}{20}$ $-\frac{13}{20}$

3.
$8\frac{6}{7}$ $-5\frac{2}{7}$ $10\frac{5}{8}$ $-4\frac{5}{8}$ $7\frac{8}{9}$ $-6\frac{5}{9}$ $13\frac{9}{10}$ $-9\frac{5}{10}$

4.
$15\frac{7}{16}$ $-7\frac{3}{16}$ $14\frac{11}{12}$ $-5\frac{5}{12}$ $18\frac{9}{13}$ $-9\frac{4}{13}$ $23\frac{5}{6}$ $-7\frac{1}{6}$

Subtracting Fractions
with Different Bottom Numbers

If the bottom numbers in a subtraction problem are different, find the LCD and raise the fractions to higher terms. Then follow the rules on **page 21.**

EXAMPLE

$$\frac{5}{8} = \frac{15}{24}$$

$$-\frac{1}{3} = \frac{8}{24}$$

$$\frac{7}{24}$$

Step 1. The LCD is $8 \times 3 = 24$.

Step 2. Raise each fraction to 24ths.

Step 3. Subtract. $15 - 8 = 7$.

Subtract and reduce.

1. $\frac{3}{4}$ $\frac{5}{8}$ $\frac{5}{6}$ $\frac{3}{4}$ $\frac{1}{2}$
   $-\frac{1}{2}$ $-\frac{1}{4}$ $-\frac{1}{3}$ $-\frac{3}{16}$ $-\frac{3}{10}$

2. $\frac{2}{3}$ $\frac{4}{5}$ $\frac{3}{4}$ $\frac{5}{6}$ $\frac{5}{9}$
   $-\frac{1}{4}$ $-\frac{1}{3}$ $-\frac{2}{7}$ $-\frac{3}{5}$ $-\frac{1}{6}$

3. $8\frac{11}{12}$ $9\frac{5}{7}$ $12\frac{4}{5}$ $11\frac{3}{4}$
   $-2\frac{3}{8}$ $-3\frac{1}{2}$ $-5\frac{2}{9}$ $-8\frac{7}{10}$

4. $13\frac{4}{7}$ $21\frac{8}{9}$ $9\frac{5}{6}$ $20\frac{9}{11}$
   $-7\frac{3}{8}$ $-6\frac{1}{4}$ $-2\frac{2}{5}$ $-8\frac{2}{3}$

5.

$7\frac{1}{2}$ $\qquad$ $12\frac{3}{4}$ $\qquad$ $5\frac{2}{3}$ $\qquad$ $4\frac{11}{16}$ $\qquad$ $18\frac{7}{10}$

$-2\frac{1}{4}$ $\qquad$ $-9\frac{3}{8}$ $\qquad$ $-3\frac{4}{7}$ $\qquad$ $-2\frac{3}{8}$ $\qquad$ $-9\frac{1}{2}$

6.

$22\frac{8}{15}$ $\qquad$ $8\frac{9}{28}$ $\qquad$ $35\frac{11}{30}$ $\qquad$ $14\frac{11}{12}$ $\qquad$ $9\frac{11}{36}$

$-17\frac{2}{9}$ $\qquad$ $-7\frac{2}{7}$ $\qquad$ $-28\frac{1}{3}$ $\qquad$ $-7\frac{5}{8}$ $\qquad$ $-5\frac{2}{9}$

7.

$15\frac{7}{8}$ $\qquad$ $45\frac{8}{9}$ $\qquad$ $5\frac{7}{8}$ $\qquad$ $10\frac{5}{6}$ $\qquad$ $100\frac{5}{8}$

$-8\frac{4}{7}$ $\qquad$ $-38\frac{5}{7}$ $\qquad$ $-4\frac{5}{9}$ $\qquad$ $-6\frac{3}{7}$ $\qquad$ $-98\frac{1}{6}$

8.

$11\frac{1}{3}$ $\qquad$ $24\frac{5}{6}$ $\qquad$ $8\frac{3}{4}$ $\qquad$ $17\frac{3}{4}$ $\qquad$ $15\frac{1}{3}$

$-5\frac{1}{8}$ $\qquad$ $-21\frac{4}{5}$ $\qquad$ $-6\frac{2}{5}$ $\qquad$ $-9\frac{2}{3}$ $\qquad$ $-7\frac{3}{10}$

9.

$12\frac{1}{2}$ $\qquad$ $13\frac{5}{8}$ $\qquad$ $16\frac{5}{6}$ $\qquad$ $18\frac{3}{5}$ $\qquad$ $11\frac{2}{3}$

$-8\frac{3}{7}$ $\qquad$ $-11\frac{5}{12}$ $\qquad$ $-9\frac{7}{10}$ $\qquad$ $-9\frac{3}{10}$ $\qquad$ $-5\frac{1}{2}$

10.

$7\frac{1}{2}$ $\qquad$ $19\frac{7}{12}$ $\qquad$ $21\frac{14}{15}$ $\qquad$ $13\frac{11}{18}$ $\qquad$ $25\frac{5}{8}$

$-3\frac{2}{5}$ $\qquad$ $-8\frac{3}{7}$ $\qquad$ $-18\frac{7}{10}$ $\qquad$ $-8\frac{1}{2}$ $\qquad$ $-22\frac{2}{5}$

Borrowing and Subtracting Fractions

In order to have a fraction to subtract from, you sometimes have to borrow from a whole number. Look at the examples carefully.

EXAMPLE 1. $9 - 6\frac{3}{5}$

Since there is nothing to subtract the $\frac{3}{5}$ from, you have to borrow.

$$9 = 8\frac{5}{5}$$
$$-6\frac{3}{5} = 6\frac{3}{5}$$
$$\overline{2\frac{2}{5}}$$

Step 1. Borrow 1 from the 9 and change the 1 to 5ths because 5 is the LCD.
$$1 = \frac{5}{5}$$

Step 2. Subtract the top numbers and the whole numbers.

EXAMPLE 2. $12\frac{3}{7} - 8\frac{6}{7}$

Since you cannot take $\frac{6}{7}$ from $\frac{3}{7}$, you have to borrow.

$$12\frac{3}{7} = 11\frac{7}{7} + \frac{3}{7}$$
$$- \ 8\frac{6}{7}$$

Step 1. Borrow 1 from 12 and change the 1 to 7ths because 7 is the LCD.
$$1 = \frac{7}{7}$$

$$11\frac{7}{7} + \frac{3}{7} = 11\frac{10}{7}$$
$$- \ 8\frac{6}{7} \qquad = \ 8\ \frac{6}{7}$$
$$\overline{3\ \frac{4}{7}}$$

Step 2. Add the $\frac{7}{7}$ to $\frac{3}{7}$. $\frac{7}{7} + \frac{3}{7} = \frac{10}{7}$

Step 3. Subtract the top numbers and the whole numbers.

EXAMPLE 3. $8\frac{1}{3} - 4\frac{3}{4}$

$$8\ \frac{1}{3} = 8\frac{4}{12}$$
$$-4\ \frac{3}{4} = 4\frac{9}{12}$$

Step 1. Raise each fraction to 12ths because 12 is the LCD.

$$8\frac{4}{12} = 7\frac{12}{12} + \frac{4}{12}$$

$$-4\frac{9}{12}$$

Step 2. Borrow 1 from 8 and change the 1 to 12ths.
$$1 = \frac{12}{12}$$

$$7\frac{12}{12} + \frac{4}{12} = 7\frac{16}{12}$$

$$-4\frac{9}{12} = 4\frac{9}{12}$$

$$3\frac{7}{12}$$

Step 3. Add the $\frac{12}{12}$ to $\frac{4}{12}$. $\frac{12}{12} + \frac{4}{12} = \frac{16}{12}$

Step 4. Subtract the top numbers and the whole numbers.

Subtract and reduce.

1.
$$8 \quad\quad 4 \quad\quad 12 \quad\quad 9 \quad\quad 10$$
$$-\frac{5}{6} \quad -\frac{3}{7} \quad -\frac{1}{2} \quad -\frac{2}{5} \quad -\frac{8}{11}$$

2.
$$12 \quad\quad 9 \quad\quad 7 \quad\quad 6 \quad\quad 10$$
$$-8\frac{3}{7} \quad -5\frac{2}{3} \quad -5\frac{7}{12} \quad -2\frac{5}{9} \quad -3\frac{5}{16}$$

3.
$$8\frac{2}{9} \quad\quad 11\frac{3}{8} \quad\quad 14\frac{7}{12} \quad\quad 15\frac{1}{5}$$
$$-4\frac{5}{9} \quad -4\frac{7}{8} \quad -6\frac{11}{12} \quad -8\frac{4}{5}$$

4.
$$12\frac{7}{15} \quad\quad 22\frac{4}{7} \quad\quad 19\frac{1}{3} \quad\quad 18\frac{3}{16}$$
$$-7\frac{8}{15} \quad -6\frac{6}{7} \quad -12\frac{2}{3} \quad -10\frac{7}{16}$$

5.
$$20\frac{1}{8} \quad\quad 36\frac{6}{11} \quad\quad 15\frac{7}{20} \quad\quad 30\frac{5}{9}$$
$$-9\frac{5}{8} \quad -8\frac{9}{11} \quad -14\frac{13}{20} \quad -27\frac{8}{9}$$

6.

$12\frac{2}{5}$ $\qquad$ $14\frac{3}{8}$ $\qquad$ $7\frac{2}{3}$ $\qquad$ $18\frac{1}{6}$

$-\ 6\frac{3}{4}$ $\qquad$ $-\ 9\frac{3}{4}$ $\qquad$ $-2\frac{8}{9}$ $\qquad$ $-\ 3\frac{3}{4}$

7.

$25\frac{1}{2}$ $\qquad$ $20\frac{5}{12}$ $\qquad$ $19\ \frac{1}{2}$ $\qquad$ $36\frac{4}{9}$

$-\ 6\frac{4}{7}$ $\qquad$ $-11\ \frac{5}{6}$ $\qquad$ $-13\frac{7}{10}$ $\qquad$ $-\ 4\frac{3}{5}$

8.

$17\frac{2}{7}$ $\qquad$ $15\frac{3}{8}$ $\qquad$ $30\ \frac{1}{3}$ $\qquad$ $12\ \frac{1}{6}$

$-\ 8\frac{5}{6}$ $\qquad$ $-\ 7\frac{4}{5}$ $\qquad$ $-16\frac{8}{11}$ $\qquad$ $-10\frac{7}{12}$

9.

$14\ \frac{1}{2}$ $\qquad$ $18\frac{4}{9}$ $\qquad$ $22\frac{5}{8}$ $\qquad$ $16\frac{7}{12}$

$-13\frac{11}{15}$ $\qquad$ $-14\frac{3}{4}$ $\qquad$ $-\ 8\frac{6}{7}$ $\qquad$ $-\ 9\ \frac{7}{8}$

10.

$28\frac{1}{6}$ $\qquad$ $15\frac{3}{4}$ $\qquad$ $30\ \frac{1}{4}$ $\qquad$ $19\frac{3}{11}$

$-17\frac{3}{5}$ $\qquad$ $-\ 8\frac{7}{8}$ $\qquad$ $-16\frac{5}{12}$ $\qquad$ $-18\ \frac{1}{2}$

11.

$17\frac{2}{7}$ $\qquad$ $35\frac{1}{4}$ $\qquad$ $13\frac{2}{9}$ $\qquad$ $24\frac{3}{16}$

$-15\frac{5}{8}$ $\qquad$ $-18\frac{3}{5}$ $\qquad$ $-\ 7\frac{5}{6}$ $\qquad$ $-\ 9\ \frac{2}{3}$

Subtraction of Fractions: Applying Your Skills

Subtraction problems ask you to find the **difference** between two numbers, to figure out what is **left** after you take something away, or to figure out an **increase** or **decrease**. Reduce all your answers to lowest terms.

1. From a board $38\frac{1}{2}$ inches long, Pete cut a piece $17\frac{5}{8}$ inches long. How long was the remaining piece?

2. Jeff weighed 166 pounds. When he was sick he lost $11\frac{3}{4}$ pounds. How much did he weigh after this loss?

3. Before leaving on a weekend trip, Mr. Green noticed that his mileage gauge registered $20{,}245\frac{3}{10}$ miles. When he returned, his mileage gauge registered $20{,}734\frac{7}{10}$ miles. How many miles did he drive that weekend?

4. Adrienne had a five-pound bag of flour. If she used $1\frac{1}{6}$ pounds of flour for a certain recipe, how much flour did she have left?

5. Shirley bought a 10-pound bag of dog food. At his first feeding, her dog ate $1\frac{3}{8}$ pounds of the food. How much dog food was left?

6. If you change a turntable speed set at 45 rpm's to $33\frac{1}{3}$ rpm's, how many revolutions per minute slower does the turntable spin?

7. From a 100-pound bag of cement, Fred used $44\frac{5}{8}$ pounds to make concrete. How much cement was left in the bag?

8. Esther had a $\frac{3}{4}$-pound bar of cooking chocolate. If she used $\frac{5}{8}$ pound of chocolate to make a cake, how much chocolate was left?

Multiplying Fractions

To multiply fractions, multiply the top numbers together and the bottom numbers together.

EXAMPLE

$$\frac{3}{5} \times \frac{4}{7} = \frac{12}{35}$$

Step 1. Multiply the top numbers. $3 \times 4 = 12$.

Step 2. Multiply the bottom numbers. $5 \times 7 = 35$.

Multiply and reduce.

1. $\frac{2}{3} \times \frac{4}{5} =$ $\frac{5}{7} \times \frac{2}{9} =$ $\frac{1}{8} \times \frac{7}{10} =$ $\frac{3}{11} \times \frac{5}{8} =$

2. $\frac{1}{3} \times \frac{1}{5} =$ $\frac{4}{7} \times \frac{4}{9} =$ $\frac{5}{6} \times \frac{5}{8} =$ $\frac{9}{10} \times \frac{1}{4} =$

3. $\frac{7}{9} \times \frac{2}{5} =$ $\frac{3}{8} \times \frac{7}{8} =$ $\frac{1}{6} \times \frac{5}{6} =$ $\frac{8}{9} \times \frac{2}{9} =$

With three fractions, multiply the top numbers of the first two fractions together. Then multiply that answer by the third top number. Do the same for the bottom numbers.

4. $\frac{3}{5} \times \frac{1}{2} \times \frac{3}{4} =$ $\frac{5}{7} \times \frac{1}{3} \times \frac{1}{2} =$ $\frac{2}{3} \times \frac{1}{3} \times \frac{5}{9} =$

5. $\frac{4}{5} \times \frac{4}{5} \times \frac{1}{3} =$ $\frac{2}{5} \times \frac{7}{9} \times \frac{1}{3} =$ $\frac{1}{3} \times \frac{4}{7} \times \frac{2}{3} =$

Canceling and Multiplying Fractions

Canceling is a shortcut in multiplication of fractions. It is just like reducing. It means dividing a top and a bottom number by a figure that goes evenly into both before actually multiplying. You don't have to cancel to get the right answer, but it makes the multiplication easier.

EXAMPLE: $\dfrac{10}{21} \times \dfrac{14}{25}$

$\dfrac{\overset{2}{\cancel{10}}}{21} \times \dfrac{14}{\underset{5}{\cancel{25}}} =$

Step 1. Cancel 10 and 25 by 5. 10 ÷ 5 = 2 and 25 ÷ 5 = 5. Cross out the 10 and the 25.

$\dfrac{\overset{2}{\cancel{10}}}{\underset{3}{\cancel{21}}} \times \dfrac{\overset{2}{\cancel{14}}}{\underset{5}{\cancel{25}}} = \dfrac{4}{15}$

Step 2. Cancel 14 and 21 by 7. 14 ÷ 7 = 2 and 21 ÷ 7 = 3. Cross out the 14 and the 21.

Step 3. Multiply across by the new numbers. 2 × 2 = 4 and 3 × 5 = 15.

Cancel and multiply.

1. $\dfrac{2}{5} \times \dfrac{3}{4} =$ $\dfrac{4}{9} \times \dfrac{3}{7} =$ $\dfrac{5}{8} \times \dfrac{7}{10} =$ $\dfrac{6}{7} \times \dfrac{5}{12} =$

2. $\dfrac{4}{5} \times \dfrac{1}{6} =$ $\dfrac{8}{15} \times \dfrac{10}{13} =$ $\dfrac{9}{14} \times \dfrac{10}{11} =$ $\dfrac{12}{13} \times \dfrac{1}{15} =$

3. $\dfrac{4}{9} \times \dfrac{3}{8} =$ $\dfrac{5}{12} \times \dfrac{9}{10} =$ $\dfrac{7}{22} \times \dfrac{11}{14} =$ $\dfrac{5}{6} \times \dfrac{9}{10} =$

4. $\dfrac{15}{16} \times \dfrac{12}{25} =$ $\dfrac{7}{24} \times \dfrac{32}{35} =$ $\dfrac{21}{26} \times \dfrac{13}{28} =$ $\dfrac{19}{45} \times \dfrac{25}{38} =$

5. $\dfrac{9}{16} \times \dfrac{8}{15} =$ $\dfrac{6}{7} \times \dfrac{28}{33} =$ $\dfrac{5}{11} \times \dfrac{22}{25} =$ $\dfrac{8}{15} \times \dfrac{9}{32} =$

6. $\dfrac{12}{13} \times \dfrac{3}{16} =$ $\dfrac{4}{9} \times \dfrac{9}{14} =$ $\dfrac{15}{16} \times \dfrac{16}{21} =$ $\dfrac{11}{24} \times \dfrac{8}{11} =$

7. $\dfrac{7}{8} \times \dfrac{9}{14} \times \dfrac{5}{6} =$ $\dfrac{4}{11} \times \dfrac{5}{12} \times \dfrac{11}{15} =$ $\dfrac{16}{21} \times \dfrac{14}{15} \times \dfrac{3}{4} =$

8. $\dfrac{9}{16} \times \dfrac{20}{21} \times \dfrac{7}{10} =$ $\dfrac{4}{15} \times \dfrac{7}{12} \times \dfrac{3}{4} =$ $\dfrac{9}{10} \times \dfrac{1}{6} \times \dfrac{5}{8} =$

9. $\dfrac{7}{24} \times \dfrac{2}{3} \times \dfrac{16}{35} =$ $\dfrac{3}{20} \times \dfrac{18}{25} \times \dfrac{5}{6} =$ $\dfrac{11}{12} \times \dfrac{5}{11} \times \dfrac{8}{15} =$

10. $\dfrac{5}{21} \times \dfrac{1}{9} \times \dfrac{15}{32} =$ $\dfrac{6}{17} \times \dfrac{21}{40} \times \dfrac{4}{45} =$ $\dfrac{19}{36} \times \dfrac{7}{10} \times \dfrac{3}{7} =$

11. $\dfrac{11}{39} \times \dfrac{10}{21} \times \dfrac{13}{18} =$ $\dfrac{17}{21} \times \dfrac{14}{51} \times \dfrac{7}{11} =$ $\dfrac{15}{28} \times \dfrac{7}{16} \times \dfrac{12}{45} =$

Multiplying Fractions by Whole Numbers

Any whole number can be written as a fraction with a bottom number of 1. For example, 5 is the same as $\frac{5}{1}$.

EXAMPLE: $9 \times \frac{5}{6}$

$$\overset{3}{\cancel{9}} \times \frac{5}{\underset{2}{\cancel{6}}} = \frac{15}{2} = 7\frac{1}{2}$$

Step 1. Write 9 as a fraction. $9 = \frac{9}{1}$.

Step 2. Cancel 9 and 6 by 3.

Step 3. Multiply across by the new numbers.

Step 4. Change the improper fraction to a mixed number (see page 11).

Multiply and reduce.

1. $4 \times \frac{3}{7} =$ $9 \times \frac{1}{4} =$ $\frac{2}{3} \times 10 =$ $3 \times \frac{4}{5} =$

2. $15 \times \frac{2}{3} =$ $\frac{5}{9} \times 18 =$ $\frac{4}{21} \times 7 =$ $\frac{8}{15} \times 45 =$

3. $\frac{7}{8} \times 24 =$ $\frac{11}{40} \times 20 =$ $32 \times \frac{7}{16} =$ $12 \times \frac{15}{16} =$

4. $35 \times \frac{7}{30} =$ $16 \times \frac{5}{24} =$ $\frac{7}{12} \times 36 =$ $2 \times \frac{9}{10} =$

Multiplying with Mixed Numbers

To multiply with mixed numbers, change every mixed number to an improper fraction. (*See page 12.*)

EXAMPLE: $4\frac{1}{2} \times \frac{5}{6} =$

$$\frac{\overset{3}{\cancel{9}}}{2} \times \frac{5}{\underset{2}{\cancel{6}}} = \frac{15}{4} = 3\frac{3}{4}$$

Step 1. Change $4\frac{1}{2}$ to an improper fraction.

$4\frac{1}{2} = \frac{9}{2}$

Step 2. Cancel 9 and 6 by 3.

Step 3. Multiply across.

Step 4. Change the improper fraction to a mixed number (see page 11).

Multiply and reduce.

1. $1\frac{1}{2} \times \frac{1}{4} =$ $1\frac{2}{3} \times \frac{2}{7} =$ $2\frac{1}{4} \times \frac{7}{8} =$ $\frac{3}{10} \times 5\frac{1}{2} =$

2. $\frac{4}{9} \times 3\frac{3}{4} =$ $\frac{2}{7} \times 2\frac{5}{8} =$ $4\frac{2}{3} \times \frac{15}{16} =$ $6\frac{3}{7} \times \frac{4}{5} =$

3. $2\frac{1}{3} \times 1\frac{1}{5} =$ $6\frac{2}{3} \times 3\frac{3}{4} =$ $3\frac{5}{7} \times 4\frac{3}{8} =$ $16\frac{1}{3} \times 2\frac{5}{14} =$

4. $3\frac{3}{4} \times \frac{8}{9} \times 1\frac{1}{5} =$ $2\frac{2}{5} \times 3\frac{3}{8} \times 2\frac{7}{9} =$ $2\frac{2}{15} \times 5\frac{1}{4} \times 7\frac{1}{2} =$

Multiplication of Fractions: Applying Your Skills

Multiplication problems with fractions generally ask you to **find a part of something** or to **figure out the cost, weight, or size of several things** when you have information about only one thing. Reduce all your answers to lowest terms.

1. Out of a 40-hour work week, Jack worked $\frac{4}{5}$ of the time. How many hours did he work?

$$\frac{\cancel{5}}{\cancel{4}} \times \frac{1}{\cancel{40}_8} = \frac{*}{32} \text{ horas}$$

2. From their house to their parents' house, the Leightons have to drive 276 miles. If they have already driven $\frac{2}{3}$ of the distance, how far have they gone?

$$\frac{2}{\cancel{3}_1} \times \frac{\cancel{276}^{92}}{1} = \frac{184}{1}$$

3. If one cubic foot of water weighs $62\frac{1}{2}$ pounds, how much do $3\frac{1}{5}$ cubic feet of water weigh?

$62\frac{1}{2}$ Reducir impropia.

$$\frac{\cancel{125}_{25}}{\cancel{2}_1} \times \frac{\cancel{16}^{8}}{\cancel{5}_1} = 200$$

4. For a bookcase, Tim wants 6 shelves each $28\frac{1}{2}$ inches long. What total length of shelving does he need?

$28\frac{1}{2}$ Reducir a impropia

$$\frac{56}{2} \times \frac{6}{1} = \frac{342}{2} \div 2 = 171$$

5. If one yard of material costs \$2.00, how much do $5\frac{1}{2}$ yards cost?

11.00

6. Robert makes \$4.00 an hour when he works overtime. How much does he make for $3\frac{1}{4}$ hours of overtime work?

13.00

7. Verva gets \$3.00 an hour where she works. For overtime she gets "time and a half" ($1\frac{1}{2}$ times her regular wage). How much does she get for one hour of overtime work?

$$3 \times 1\frac{1}{2} = \frac{3}{1} \times \frac{3}{2} = \frac{9}{2} \overline{)\frac{2}{4}} = \left(4\frac{1}{2}\right) = 9.00$$

8. A tailor needs $3\frac{1}{6}$ yards of material to make a suit. How much material does he need to make three suits?

$$3\frac{1}{6} \times \frac{3}{1} = \frac{19}{\cancel{6}_2} \times \frac{\cancel{3}^1}{1} = \frac{19}{2} \overline{)\frac{2}{9}} = 9\frac{1}{2}$$

Dividing Fractions by Fractions

Suppose a man owned a $\frac{1}{2}$-acre piece of land that he wanted to divide into $\frac{1}{8}$-acre sections for resale. How many $\frac{1}{8}$-acre sections will he have to sell? To answer this question, you have to find out how many $\frac{1}{8}$s are contained in $\frac{1}{2}$.

You know from the work that you've already done with fractions that 1 whole contains $\frac{8}{8}$; therefore, there are four $\frac{1}{8}$-acre sections in $\frac{1}{2}$ acre.

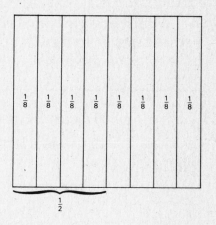

To calculate the answer to any division of fractions problem, there are two rules to remember:

(1) **Invert** the fraction to the right of the division sign (the divisor). That is, turn the fraction upside down by writing the top number in the bottom position and the bottom number at the top.

In the problem above ($\frac{1}{2} \div \frac{1}{8}$), invert the $\frac{1}{8}$ to become $\frac{8}{1}$.

(2) Change the division sign to a multiplication sign and follow the rules of multiplication.

Thus, the problem above becomes:

$$\frac{1}{2} \div \frac{1}{8} = \frac{1}{\cancel{2}} \times \frac{\cancelto{4}{8}}{1} = 4 \text{ sections}$$

In other words, the rules for multiplication and division of fractions are exactly the same as soon as you **invert the fraction to the right of the division sign.**

EXAMPLE: $\quad \frac{3}{4} \div \frac{5}{8}$

$$\frac{3}{\cancelto{1}{4}} \times \frac{\cancelto{2}{8}}{5} = \frac{6}{5} = 1\frac{1}{5}$$

Step 1. Invert the fraction on the right ($\frac{5}{8}$) to $\frac{8}{5}$ and change the $\div$ sign to $\times$.

Step 2. Cancel 4 and 8 by 4.

Step 3. Multiply across.

Step 4. Change the improper fraction to a mixed number (see page 11).

Divide and reduce.

1. $\frac{3}{7} \div \frac{2}{5} =$ $\qquad$ $\frac{4}{9} \div \frac{2}{3} =$ $\qquad$ $\frac{5}{12} \div \frac{10}{11} =$ $\qquad$ $\frac{7}{15} \div \frac{4}{5} =$

2. $\frac{11}{16} \div \frac{5}{8} =$ $\qquad$ $\frac{8}{9} \div \frac{2}{9} =$ $\qquad$ $\frac{8}{13} \div \frac{6}{7} =$ $\qquad$ $\frac{21}{25} \div \frac{7}{10} =$

3. $\frac{14}{15} \div \frac{16}{27} =$ $\qquad$ $\frac{35}{36} \div \frac{25}{28} =$ $\qquad$ $\frac{9}{14} \div \frac{1}{2} =$ $\qquad$ $\frac{6}{25} \div \frac{1}{5} =$

4. $\frac{1}{12} \div \frac{7}{9} =$ $\qquad$ $\frac{4}{11} \div \frac{1}{11} =$ $\qquad$ $\frac{49}{50} \div \frac{21}{65} =$ $\qquad$ $\frac{8}{9} \div \frac{5}{12} =$

5. $\frac{5}{36} \div \frac{11}{42} =$ $\qquad$ $\frac{13}{63} \div \frac{7}{54} =$ $\qquad$ $\frac{17}{48} \div \frac{1}{24} =$ $\qquad$ $\frac{1}{18} \div \frac{8}{27} =$

6. $\frac{3}{10} \div \frac{6}{7} =$ $\qquad$ $\frac{5}{11} \div \frac{25}{33} =$ $\qquad$ $\frac{12}{19} \div \frac{18}{38} =$ $\qquad$ $\frac{16}{21} \div \frac{3}{4} =$

Dividing Whole Numbers by Fractions

Suppose that the owner of the $\frac{1}{2}$-acre piece of land (page 34) has another 3 acres of land that he wants to divide into $\frac{3}{4}$-acre sections for resale. How many $\frac{3}{4}$-acre sections will he have to sell? To answer this question, you have to find out how many $\frac{3}{4}$s are contained in 3 acres.

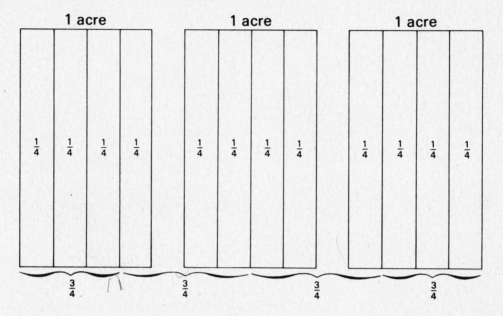

From the illustration, you can see that he will have four $\frac{3}{4}$-acre sections to sell.

When dividing whole numbers by fractions, write the whole number as a fraction by putting it over 1 $\left(\frac{3}{1} \div \frac{3}{4}\right)$, invert the fraction to the right of the division sign and multiply:

$$\frac{\cancel{3}^{1}}{1} \times \frac{4}{\cancel{3}_{1}} = 4 \text{ sections}$$

EXAMPLE: $8 \div \frac{6}{7}$

$$\frac{\cancel{8}^{4}}{1} \times \frac{7}{\cancel{6}_{3}} = \frac{28}{3} = 9\frac{1}{3}$$

Step 1.   Write 8 as a fraction: $\frac{8}{1}$

Step 2.   Invert $\frac{6}{7}$ and change the $\div$ sign to $\times$.

Step 3.   Cancel 8 and 6 by 2.

Step 4.   Multiply across.

Step 5.   Change the improper fraction to a mixed number.

Divide and reduce.

1. $12 \div \frac{2}{5} =$ $\qquad$ $9 \div \frac{1}{3} =$ $\qquad$ $5 \div \frac{3}{4} =$ $\qquad$ $6 \div \frac{2}{3} =$

2. $8 \div \frac{3}{8} =$ $\qquad$ $25 \div \frac{5}{4} =$ $\qquad$ $17 \div \frac{1}{2} =$ $\qquad$ $32 \div \frac{8}{9} =$

3. $45 \div \frac{9}{10} =$ $\qquad$ $49 \div \frac{7}{12} =$ $\qquad$ $36 \div \frac{24}{25} =$ $\qquad$ $54 \div \frac{9}{11} =$

4. $48 \div \frac{8}{15} =$ $\qquad$ $27 \div \frac{3}{4} =$ $\qquad$ $16 \div \frac{3}{5} =$ $\qquad$ $30 \div \frac{20}{21} =$

5. $32 \div \frac{4}{21} =$ $\qquad$ $45 \div \frac{18}{19} =$ $\qquad$ $15 \div \frac{10}{11} =$ $\qquad$ $12 \div \frac{9}{10} =$

6. $20 \div \frac{5}{8} =$ $\qquad$ $5 \div \frac{15}{16} =$ $\qquad$ $7 \div \frac{21}{25} =$ $\qquad$ $56 \div \frac{24}{25} =$

Dividing Fractions by Whole Numbers

Suppose that you've eaten 3 slices of an 8-slice pizza. This leaves $\frac{5}{8}$ of the pizza. A friend joins you, and you want to divide the rest of the pizza equally between you. How much pizza would each person get?

To answer this question, you will have to divide the fraction $\left(\frac{5}{8}\right)$ by the number of people (2) that want to share the pizza.

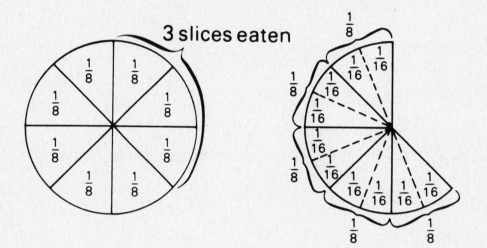

3 slices eaten

Notice that each piece is divided into two equal parts. $\frac{1}{2}$ of $\frac{1}{8}$ is $\frac{1}{16}$, making $\frac{10}{16}$. Half of $\frac{10}{16}$ is $\frac{5}{16}$.

When dividing a fraction by a whole number, first write the whole number as a fraction over 1 $\left(\frac{5}{8} \div \frac{2}{1}\right)$. Then invert that fraction and multiply:

$$\frac{5}{8} \times \frac{1}{2} = \frac{5}{16}$$

EXAMPLE: $\frac{3}{4} \div 6$

$\frac{3}{4} \div \frac{6}{1} =$ *Step 1.* Write 6 as a fraction: $\frac{6}{1}$

$\frac{\cancel{3}}{4} \times \frac{1}{\cancel{6}_2} = \frac{1}{8}$ *Step 2.* Invert the fraction $\frac{6}{1}$ to $\frac{1}{6}$ and change the ÷ sign to ×.

Step 3. Cancel 3 and 6 by 3.

Step 4. Multiply across.

Divide and reduce.

1. $\frac{4}{5} \div 4 =$ $\frac{2}{3} \div 6 =$ $\frac{18}{19} \div 9 =$ $\frac{1}{4} \div 2 =$

2. $\frac{1}{3} \div 12 =$ $\frac{3}{4} \div 7 =$ $\frac{5}{6} \div 10 =$ $\frac{12}{13} \div 3 =$

3. $\frac{1}{2} \div 11 =$ $\frac{3}{5} \div 15 =$ $\frac{5}{8} \div 5 =$ $\frac{12}{13} \div 18 =$

4. $\frac{20}{21} \div 24 =$ $\frac{9}{10} \div 36 =$ $\frac{14}{15} \div 35 =$ $\frac{24}{25} \div 40 =$

5. $\frac{3}{5} \div 9 =$ $\frac{10}{13} \div 30 =$ $\frac{12}{17} \div 36 =$ $\frac{21}{22} \div 28 =$

6. $\frac{4}{7} \div 44 =$ $\frac{15}{16} \div 45 =$ $\frac{5}{18} \div 15 =$ $\frac{7}{24} \div 35 =$

Dividing with Mixed Numbers

To divide with mixed numbers, first change every mixed number to an improper fraction (see page 12). Also, be sure to write whole numbers as fractions over 1. Then invert the fraction to the right of the division sign and finish the problems as on pages 34, 36, and 38.

EXAMPLE: $2\frac{1}{3} \div \frac{1}{4}$

$$\frac{7}{3} \div \frac{1}{4}$$

$$\frac{7}{3} \times \frac{4}{1} = \frac{28}{3} = 9\frac{1}{3}$$

Step 1. Change $2\frac{1}{3}$ to an improper fraction: $\frac{7}{3}$

Step 2. Invert the fraction $\frac{1}{4}$ to $\frac{4}{1}$ and change the $\div$ sign to $\times$.

Step 3. Since nothing can be canceled, multiply across.

Step 4. Change the improper fraction to a mixed number.

Divide and reduce.

1. $1\frac{1}{2} \div \frac{3}{4} =$ $1\frac{2}{3} \div \frac{2}{3} =$ $2\frac{3}{4} \div \frac{5}{8} =$ $4\frac{1}{3} \div \frac{2}{9} =$

2. $2\frac{2}{5} \div 6 =$ $3\frac{1}{3} \div 4 =$ $1\frac{5}{7} \div 9 =$ $2\frac{2}{9} \div 15 =$

3. $\frac{5}{8} \div 1\frac{1}{4} =$ $\frac{14}{15} \div 1\frac{1}{6} =$ $\frac{7}{12} \div 2\frac{1}{2} =$ $\frac{9}{16} \div 3\frac{3}{4} =$

4. $12 \div 1\frac{3}{5} =$ $20 \div 2\frac{2}{7} =$ $9 \div 1\frac{7}{8} =$ $24 \div 1\frac{7}{11} =$

5. $3\frac{3}{4} \div 1\frac{1}{8} =$ $4\frac{1}{2} \div 1\frac{1}{6} =$ $2\frac{3}{4} \div 1\frac{7}{8} =$ $1\frac{7}{9} \div 2\frac{2}{9} =$

6. $5\frac{1}{4} \div 4\frac{2}{3} =$ $6\frac{1}{2} \div 3\frac{1}{4} =$ $5\frac{3}{5} \div 3\frac{3}{7} =$ $4\frac{3}{8} \div 1\frac{9}{16} =$

7. $5\frac{2}{3} \div 1\frac{8}{9} =$ $4\frac{2}{5} \div 8\frac{4}{5} =$ $3\frac{5}{9} \div 1\frac{13}{15} =$ $10\frac{2}{3} \div 2\frac{2}{3} =$

8. $3\frac{3}{5} \div 1\frac{7}{20} =$ $6\frac{3}{5} \div 2\frac{1}{5} =$ $3\frac{3}{7} \div 1\frac{11}{21} =$ $2\frac{4}{13} \div 2\frac{1}{4} =$

9. $4\frac{1}{3} \div 2\frac{1}{7} =$ $6\frac{7}{8} \div 5\frac{1}{4} =$ $9\frac{7}{10} \div 1\frac{4}{5} =$ $8\frac{2}{3} \div 5\frac{1}{12} =$

10. $7\frac{1}{2} \div 3\frac{1}{5} =$ $3\frac{5}{9} \div 2\frac{5}{18} =$ $5\frac{5}{6} \div 3\frac{5}{12} =$ $10\frac{5}{8} \div 4\frac{1}{2} =$

Division of Fractions: Applying Your Skills

Problems with division of fractions are tricky because you have to be sure that you invert the right number. An important hint to remember when you are setting up the problem is that the thing being divided or cut or shared or split should be written to the **left** of the division sign. For example, if $5\frac{1}{2}$ pounds of sugar is going to be divided evenly among 5 people, the $5\frac{1}{2}$ should be written first, then the division sign, then the 5: $5\frac{1}{2} \div 5$. The number on the **right** is always the one that gets inverted.

Reduce all your answers to lowest terms.

1. How many pieces of wood each $7\frac{1}{2}$ inches long can be cut from a board that is 45 inches long?

2. Sandy baked $4\frac{1}{2}$ pounds of cookies. If she divides the cookies evenly among herself and two friends, how many pounds of cookies will each person get?

3. If a tailor needs $3\frac{2}{3}$ yards of material to make a suit, how many suits can he make from 22 yards of material?

4. How many $1\frac{1}{2}$-pound loaves of bread can be made from 9 pounds of dough?

5. How many $\frac{3}{4}$-pound cans of tomatoes can be filled from 24 pounds of tomatoes?

6. A piece of moulding 75 inches long is to be cut into small strips each $8\frac{1}{3}$ inches long to make picture frames. How many strips can be cut from the long piece?

7. Sarah needs $2\frac{1}{4}$ yards of material to make a dress for her daughter. How many dresses can she make from $10\frac{1}{2}$ yards of material?

Final Fraction Skills Inventory

Write each answer in lowest terms.

1. A yard contains 36 inches. 21 inches is what fraction of a yard?

2. Reduce $\frac{18}{32}$

3. Change $\frac{50}{12}$ to a mixed number.

4. Change $11\frac{5}{8}$ to an improper fraction.

5.
$$9\frac{7}{8}$$
$$6\frac{5}{8}$$
$$+3\frac{3}{8}$$

6.
$$4\frac{3}{5}$$
$$8\frac{1}{2}$$
$$+7\frac{2}{3}$$

7. Mr. Guttierez usually takes $\frac{3}{4}$ of an hour to drive home from work. Because of a traffic jam, he took an extra $1\frac{2}{3}$ hours to get home one night. How long did that ride take him?

8. Find the combined weight of three packages that weigh $5\frac{1}{2}$ pounds, $4\frac{7}{16}$ pounds, and $3\frac{3}{8}$ pounds.

9. $10\frac{7}{8}$
    $-\ 5\frac{2}{5}$

10. $15\frac{2}{11}$
    $-\ 9\frac{7}{11}$

11. $13\frac{2}{9}$
    $-\ 4\frac{5}{6}$

12. From a two-pound box of chocolates, Rachel ate $1\frac{1}{4}$ pounds of the chocolate. What was the weight of the remaining chocolates?

13. The distance from Ellen's home to her school is $4\frac{1}{3}$ miles. If she has already traveled $2\frac{1}{2}$ miles, how far does she have to go?

14. $\frac{5}{9} \times \frac{4}{7} =$

15. $\frac{6}{7} \times \frac{14}{15} \times \frac{1}{2} =$

16. $2\frac{1}{7} \times 1\frac{5}{9} =$

17. What is the total weight of four cartons if each one weighs $16\frac{1}{4}$ pounds?

18. If one foot of lumber costs 18¢, how much do $5\frac{1}{3}$ feet of the lumber cost?

19. $\frac{8}{15} \div \frac{7}{12} =$

20. $14 \div \frac{4}{9} =$

21. $\frac{6}{13} \div 12 =$

22. $3\frac{1}{8} \div 5\frac{5}{6} =$

23. If a carpenter needs $3\frac{1}{2}$ yards of lumber to build a bookcase, how many bookcases can he build from 21 yards of lumber?

24. For a $7\frac{1}{2}$ hour day, Ed makes $30. How much did he make in one hour?

FINAL FRACTION SKILLS INVENTORY CHART

A passing score is 20 problems correct. If you had less than 20 correct, review the fraction section before going on to page 46. Even if you had a passing score or better, any problem you missed should be corrected. Following is a list of the problems and the pages where each problem is covered.

Problem Number	Practice Page	Problem Number	Practice Page
1	6	13	27
2	8–9	14	28
3	11	15	29–30
4	12	16	32
5	13–15	17	33
6	16–19	18	33
7	20	19	34–35
8	20	20	36–37
9	22–23	21	38–39
10	24–26	22	40–41
11	24–26	23	42
12	27	24	42

Decimal Skills Inventory

Do all of the following problems that you can. There is no time limit. Work accurately, but do not use outside help.

1. Write seventy-six thousandths as a decimal.

2. Change 3.04 to a mixed number and reduce.

3. Change $\frac{5}{12}$ to a decimal.

4. Which of the following is larger — .076 or .08?

5. Rewrite the following list in order from smallest to largest: .087, .7, .08, .07

6. 4.3 + .079 + .06 =

7. 21 + 2.86 + .093 =

8. Find the total weight of three packages that weigh 2.3 pounds, 4.15 pounds, and 5.65 pounds.

9. At 5:00 on a June morning the temperature was 62.4°. By 2:00 in the afternoon the temperature had risen 21.7°. What was the temperature at 2:00 that afternoon?

10. $9.6 - .457 =$ 　　　　　　　**11.** $12 - .37 =$

12. From a piece of yarn 3 yards long, Sylvia cut off a piece 1.85 yards long. How long was the remaining piece?

13. When Fred began driving Saturday morning, his mileage gauge read 8249.6 miles. When he returned home that night, it read 8536.4 miles. How far did Fred drive that day?

14. 　　2.07
　　$\times$　5.3

15. 　　.038
　　$\times$　.6

16. $4.92 \times 10 =$

17. Diane drove her car at an average speed of 45 miles per hour for 3.2 hours. How far did she drive?

18. At $3.60 an hour, how much would a worker make in 7.25 hours?

19. $14\overline{)12.04}$ 　　　　**20.** $3.2\overline{)23.36}$ 　　　　**21.** $.015\overline{)5.55}$

22. $.075\overline{)15}$

23. $49.6 \div 100 =$

24. If 1.4 pounds of meat cost $2.52, how much does one pound of the meat cost?

25. One yard contains 0.914 meters. How many yards are there in 4.57 meters?

DECIMAL SKILLS INVENTORY CHART

A passing score is 20 problems correct. If you had less than 20 correct, complete pages 49 through 72 before going on. Even if you had a passing score or better, any problem you missed should be corrected. Following is a list of the problems and the pages where each problem is covered.

Problem Number	Practice Page	Problem Number	Practice Page
1	51	14	59–61
2	52	15	59–61
3	53	16	62
4	54	17	63
5	54	18	63
6	55	19	64
7	55	20	65–66
8	56	21	65–66
9	56	22	67
10	57	23	68
11	57	24	69
12	58	25	69
13	58		

What Are Decimals?

Decimals are a type of fraction that you probably work with every day of your life. The following figures contain decimal fractions: $2.**75**, $8.**08**, $.**20**. As you already know they stand for dollars and cents, and the point separates the dollars from the change. But, have you ever thought about why cents are always written to the right of a point?

Since there are 100 cents in a dollar, cents are 100ths of a dollar. In the figures above seventy-five pennies is $\frac{75}{100}$ of a dollar; eight pennies is $\frac{8}{100}$ of a dollar; twenty pennies is $\frac{20}{100}$ of a dollar. The decimal point tells you that everything to the right of it is a part or a fraction of a dollar.

In our money system, decimal fractions only go up to hundredths; however, decimal fractions can go from tenths to millionths when they are used for certain types of exact measurement. Since decimal fractions don't have bottom numbers, you read them by noting the number of **places** they take up to the right of the decimal point.

Decimals are arranged by 10's only—10ths, 100ths, 1,000ths, 10,000ths, 100,000ths, 1,000,000ths.

Number of Places	Decimal Names	Examples	Proper Fraction
one place	= tenths	.4	= $\frac{4}{10}$
two places	= hundredths	.19	= $\frac{19}{100}$
three places	= thousandths	.005	= $\frac{5}{1,000}$
four places	= ten-thousandths	.0067	= $\frac{67}{10,000}$
five places	= hundred-thousandths	.00183	= $\frac{183}{100,000}$
six places	= millionths	.000072	= $\frac{72}{1,000,000}$

Mixed decimals are written with whole numbers to the left of the decimal point and decimal fractions to the right.

Mixed Decimals		Mixed Number	Read
3.4	=	$3\frac{4}{10}$	three and four tenths
16.07	=	$16\frac{7}{100}$	sixteen and seven hundredths
9.006	=	$9\frac{6}{1,000}$	nine and six thousandths
340.0012	=	$340\frac{12}{10,000}$	three hundred forty and twelve ten-thousandths

Reading Decimals

You can use the charts on page 49 to help you read any decimal. First, find the decimal point. Any number to the *left* of the decimal point is read as a whole number. Any number to the *right* of the decimal point is a decimal fraction. The number of places to the right of the decimal point will tell you the decimal name of that fraction. Remember to use the word *and* to separate the whole number from the decimal fraction.

> **EXAMPLE 1.** Read this decimal: .0042
>
> *Step 1.* Read the number: forty-two.
> *Step 2.* Count the number of places: .0042 has four places. Four places means ten-thousandths.
> *Step 3.* Read .0042 as **forty-two ten-thousandths.**

> **EXAMPLE 2.** Read this mixed decimal: 7.003
>
> *Step 1.* Read the whole number: seven
> *Step 2.* Read the decimal fraction number: three
> *Step 3.* Count the number of places the decimal fraction is to the right of the decimal point: .003 has three places. Three places means thousandths.
> *Step 4.* Read the whole number first and then the decimal fraction, inserting the word *and* between them. Thus, 7.003 is read: **seven and three thousandths.**

Write the following decimals in words.

1. .8_____   .3_____

2. .06_____   .37_____

3. .09_____   .005_____

4. .023_____   .0028_____

5. .378_____   .0006_____

6. .0041_____   .3577_____

7. 28.16_____   56.025_____

8. 4.008_____   10.375_____

9. 63.0078_____   19.00028_____

10. 8.000326_____   2.000003_____

Writing Decimals

When writing decimals from words, be sure that you have the correct number of places, and watch for the word *and*.

EXAMPLE 1.   Write eight hundredths.

Step 1.   Write the number 8.
Step 2.   Hundredths means two places. Since 8 uses only one place, we hold the first place to the right of the decimal point with a zero.

Answer: .08

EXAMPLE 2.   Write two hundred eight millionths.

Step 1.   Write the number 208.
Step 2.   Millionths means six places. Since 208 uses only three places, we hold the first three places to the right of the decimal point with zeros.

Answer: .000208

EXAMPLE 3.   Write sixteen and thirteen thousandths.

Step 1.   Write the whole number 16, and place a decimal point after it for the word *and*.
Step 2.   Write the number 13.
Step 3.   Thousandths means three places. Since 13 uses only two places, we hold the first place to the right of the decimal point with a zero.

Answer: 16.013

Write the following as decimals or mixed decimals.

1. seven tenths_____   twelve and one hundred three thousandths_____

2. four and nine tenths_____   two hundred fifty and six tenths_____

3. six thousandths_____   three hundred twenty-one ten-thousandths_____

4. twenty-two thousandths_____   three hundred ninety-nine and five tenths_____

5. fifteen ten-thousandths_____   sixty and three hundred twelve thousandths_____

6. eighty-five ten-thousandths_____   twenty and nine hundred-thousandths_____

7. four and seven hundredths_____   three hundred eighteen hundred-thousandths_____

8. nineteen millionths_____   ten thousand and ten ten-thousandths_____

Changing Decimals to Fractions

To change a decimal to a fraction (or a mixed decimal to a mixed number), write the figures in the decimal fraction as the top number and write the bottom number according to the number of places used. If you can, reduce the fraction.

EXAMPLE 1. Change .24 to a common fraction.

Step 1. Write 24 as the top number. $\dfrac{24}{}$

Step 2. Two places means hundredths. Write 100 as the bottom number. $\dfrac{24}{100}$

Step 3. Reduce the fraction. 24 and 100 can be divided evenly by 4. $\dfrac{24 \div 4}{100 \div 4} = \dfrac{6}{25}$

EXAMPLE 2. Change 9.015 to a mixed number.

Step 1. Write 9 as the whole number and 15 as the top number of the fraction. $9\dfrac{15}{}$

Step 2. Three places means thousandths. Write 1,000 as the bottom number. $9\dfrac{15}{1,000}$

Step 3. Reduce the fraction. 15 and 1,000 can be divided by 5. $9\dfrac{15 \div 5}{1,000 \div 5} = 9\dfrac{3}{200}$

Write each of the following as a common fraction or a mixed number and reduce.

1. .08 =   .375 =   .0048 =

2. 3.6 =   9.86 =   10.002 =

3. .085 =   5.08 =   .0025 =

4. .00324 =   19.0786 =   123.462 =

5. 16.00004 =   7.22 =   3.000008 =

6. 2036.8 =   48.02 =   3.075 =

Changing Fractions to Decimals

To change a fraction to a decimal, divide the bottom number into the top number. To do this, add a decimal point and zeros to the top number. Usually, two zeros are enough. Bring the point up into the answer.

EXAMPLE 1. Change $\frac{1}{2}$ to a decimal.

Step 1. Divide the bottom number (2) into the top number (1).

Step 2. Add a decimal point and zeros. Divide and bring the point up.

NOTE: Here, one zero was enough to complete the division problem.

$$\begin{array}{r} .5 \\ 2\overline{)1.0} \\ 1\,0 \\ \hline 0 \end{array}$$

EXAMPLE 2. Change $\frac{3}{20}$ to a decimal.

Step 1. Divide the bottom number into the top.

Step 2. Add a decimal point and zeros. Divide and bring the point up.

$$\begin{array}{r} .15 \\ 20\overline{)3.00} \\ 2\,0 \\ \hline 1\,00 \\ 1\,00 \\ \hline 0 \end{array}$$

EXAMPLE 3. Change $\frac{2}{3}$ to a decimal.

Step 1. Divide the bottom number into the top.

Step 2. Add a decimal point and zeros. Divide and bring the point up.

NOTE: Here, the division will not come out evenly no matter how many zeros you add. After two places, you write the remainder as a fraction over the number you divided by.

$$\begin{array}{r} .66\frac{2}{3} \\ 3\overline{)2.00} \\ 1\,8 \\ \hline 20 \\ 18 \\ \hline 2 \end{array}$$

Change each of the following to decimals.

1. $\frac{1}{4} =$ $\frac{2}{5} =$ $\frac{5}{8} =$ $\frac{1}{3} =$

2. $\frac{2}{9} =$ $\frac{6}{25} =$ $\frac{1}{6} =$ $\frac{3}{8} =$

3. $\frac{5}{6} =$ $\frac{3}{10} =$ $\frac{4}{7} =$ $\frac{5}{12} =$

Comparing Decimals

When you look at a group of decimal fractions, it is sometimes difficult to tell which one is the largest. Here is a trick you can use to compare decimals.

To compare decimals, give each decimal you are comparing the same number of places by adding zeros. This is the same as giving each decimal a common denominator.

EXAMPLE 1. Which of the following is larger — .07 or .2?

Step 1. Add one zero to .2

By adding the zero, we have .07 and .20

Step 2. Compare

Since twenty hundredths is larger than seven hundredths, the answer is .2

EXAMPLE 2. Arrange the following decimals in order from the smallest to the largest: .8, .08, .088, .808

Step 1. Add zeros so that each decimal in the problem has the same denominator

Adding zeros, we have .800, .080, .088, .808

Step 2. Compare and arrange the decimals in the correct order

In order, we have .08, .088, .8, .808

Notice that the extra zeros are not written in the final list.

Tell which decimal is larger in each of the following pairs.

1. .04 or .008

2. .9 or .99

3. .328 or .33

4. .0792 or .11

5. .0057 or .006

6. .4 or .0444

Arrange each of the following lists in order from the smallest to the largest.

7. .03, .33, .033, .303

8. .082, .28, .8, .08

9. .106, .16, .061, .6

10. .017, .2, .02, .007

11. .4, .405, .45, .045

12. .04, .304, .32, .4

13. .0072, .07, .027, .02

14. .2, .06, .0602, .026

Adding Decimals

To add decimals, first line them up with *point under point*. **Remember:** Any whole number is understood to have a decimal point at its right.

EXAMPLE: Add 2.46 + .005 + 16

Step 1.   Line up *point under point*. Notice the decimal point after the whole number 16.

Step 2.   Add.

$$\begin{array}{r} 2.46 \\ .005 \\ +16. \\ \hline 18.465 \end{array}$$

Add the following.

1. .8 + .047 + .36 =

2. 4.9 + 17 + 3.28 =

3. 123 + 2.6 + 9.04 =

4. 32.637 + 5 + 1.98 =

5. 9.043 + .27 + 15 =

6. 8.04 + 26 + 31.263 =

7. .849 + 1.6 + 73 =

8. .0097 + 2.8 + 16 =

9. 7.563 + .08 + 124.9 =

10. 83.007 + .47 + 9.8 =

Addition of Decimals: Applying Your Skills

1. The average monthly rainfall in New York City is 2.96 inches in June, 3.69 inches in July, and 4.01 inches in August. What is the total rainfall for these three months?

2. Thursday Jack drove 278.5 miles; Friday he drove 243.7 miles; Saturday he drove 386 miles; and Sunday he drove 291.8 miles. What total distance did he drive those four days?

3. Aldo welded together pieces of pipe that were 25.6 inches long, 19.8 inches long, and 31.5 inches long. How long was the pipe made of the three welded parts?

4. Dorothy bought 2.6 pounds of beef, 1.75 pounds of cheese, 3 pounds of chicken, and 2.35 pounds of fish. What was the total weight she had to carry?

5. The distance from Middletown to Springfield is 72.6 miles. The distance from Springfield to Centerville is 48.9 miles. What is the distance from Middletown to Centerville by driving through Springfield?

6. Maceo's normal temperature is 98.6°. While he had a fever, his temperature went up 4.5°. What was his temperature when he had a fever?

7. Find the total weight of three packages that weigh 4.2 kilograms, 2.37 kilograms, and .45 kilograms.

8. In 1970 there were 204.9 million people in the U.S. By 1973 there were 5.5 million more people. How many people were there in the U.S. in 1973?

Subtracting Decimals

To subtract decimals: put the larger number on top; line up the decimal points, add zeros to the right so that each decimal has the same number of places; and subtract as you would for whole numbers, bringing down the decimal point.

EXAMPLE: 15.2 − .184

    Step 1. Put the larger number on top and line up the decimal points.

    Step 2. Add zeros to give the top number the same number of places as the bottom number.

    Step 3. Subtract and bring down the decimal point.

$$\begin{array}{r} 15.200 \\ -.184 \\ \hline 15.016 \end{array}$$

Subtract the following.

1. 4.2 − 3.76 =

2. .804 − .1673 =

3. 3.2 − 2.68 =

4. .2 − .078 =

5. 60.4 − 6.04 =

6. 89.3 − .766 =

7. 12 − .936 =

8. 1 − .047 =

9. 13 − .932 =

10. 8.4 − .631 =

11. .07 − .002 =

12. 5 − 2.493 =

13. 96 − 7.485 =

14. 3.2 − .1986 =

15. .47 − .3992 =

Subtraction of Decimals: Applying Your Skills

1. The average American man lives for about 67.4 years. The average American woman lives for about 75.2 years. How much longer do women live?

2. In 1970 there were 792.8 people for every square mile in Puerto Rico. In 1960 there were 686.4 people for every square mile. Find the increase in the population for every square mile in Puerto Rico from 1960 to 1970.

3. From a board that was 2 meters long, Colin cut off a piece 1.67 meters long. How long was the piece he had left?

4. Fred planned to use the Interstate highway to drive on his vacation, which would mean a drive of 91.4 miles. A friend told him that the state highway was a nicer drive and would mean driving 94.5 miles. How many miles are added to the drive by taking the state highway?

5. When Sam began driving on Monday morning his mileage gauge read 16,428.7 miles. When he stopped driving Monday night, it read 16,831.4 miles. How many miles did he drive that day?

6. The metal platform of a bridge that is 124.2 meters long in the summer shrinks by 1.05 meters in the winter. How long is the bridge platform in the winter?

7. In 1965 there were 93.7 million telephones in use in the U.S. In 1970 there were 120.2 million phones. What was the increase in the number of telephones in use from 1965 to 1970?

8. If the price of a gallon of gasoline rises from 64.8¢ to 71.2¢, how much does the price of a gallon rise?

Multiplying Decimals

To multiply decimals, multiply the two numbers the same way you would whole numbers. Then count the number of decimal places in both numbers you are multiplying. Decimal places are numbers to the right of the decimal point. Put the total number of places in your answer.

EXAMPLE

$$\begin{array}{r} 4.36 \\ \times\ 2 \\ \hline 8.72 \end{array}$$
4.36   two decimal places
× 2   no decimal places
8.72   two decimal places

Multiply the following.

1.
$$\begin{array}{r} 3.8 \\ \times\ 4 \\ \hline \end{array}$$
$$\begin{array}{r} .92 \\ \times\ 9 \\ \hline \end{array}$$
$$\begin{array}{r} 6.7 \\ \times\ 6 \\ \hline \end{array}$$
$$\begin{array}{r} 5.3 \\ \times\ 8 \\ \hline \end{array}$$
$$\begin{array}{r} .84 \\ \times\ 7 \\ \hline \end{array}$$

2.
$$\begin{array}{r} 41 \\ \times.03 \\ \hline \end{array}$$
$$\begin{array}{r} 78 \\ \times.5 \\ \hline \end{array}$$
$$\begin{array}{r} 59 \\ \times.09 \\ \hline \end{array}$$
$$\begin{array}{r} 86 \\ \times.4 \\ \hline \end{array}$$
$$\begin{array}{r} 19 \\ \times.06 \\ \hline \end{array}$$

3.
$$\begin{array}{r} 34.7 \\ \times\ 8 \\ \hline \end{array}$$
$$\begin{array}{r} 2.89 \\ \times\ 7 \\ \hline \end{array}$$
$$\begin{array}{r} .551 \\ \times\ 6 \\ \hline \end{array}$$
$$\begin{array}{r} 60.3 \\ \times\ 9 \\ \hline \end{array}$$
$$\begin{array}{r} 7.18 \\ \times\ 4 \\ \hline \end{array}$$

4.
$$\begin{array}{r} 906 \\ \times.07 \\ \hline \end{array}$$
$$\begin{array}{r} 504 \\ \times.002 \\ \hline \end{array}$$
$$\begin{array}{r} 783 \\ \times.8 \\ \hline \end{array}$$
$$\begin{array}{r} 652 \\ \times.06 \\ \hline \end{array}$$
$$\begin{array}{r} 467 \\ \times.003 \\ \hline \end{array}$$

5.
$$\begin{array}{r} 2.85 \\ \times\ 50 \\ \hline \end{array}$$
$$\begin{array}{r} .693 \\ \times\ 40 \\ \hline \end{array}$$
$$\begin{array}{r} 44.7 \\ \times\ 30 \\ \hline \end{array}$$
$$\begin{array}{r} 8.01 \\ \times\ 70 \\ \hline \end{array}$$
$$\begin{array}{r} 9.17 \\ \times\ 60 \\ \hline \end{array}$$

You may need to add zeros in front of your answer to have enough decimal places in the final answer.

EXAMPLE

.06	two decimal places
× .4	one decimal place
(24	*which needs one zero to make three places*)
.024	three decimal places

6.
$$\begin{array}{r} .09 \\ \times\ .6 \\ \hline \end{array} \quad \begin{array}{r} .05 \\ \times\ .7 \\ \hline \end{array} \quad \begin{array}{r} .004 \\ \times\ \ .3 \\ \hline \end{array} \quad \begin{array}{r} .08 \\ \times.04 \\ \hline \end{array} \quad \begin{array}{r} .002 \\ \times\ .8 \\ \hline \end{array}$$

7.
$$\begin{array}{r} 5.6 \\ \times\ .2 \\ \hline \end{array} \quad \begin{array}{r} .73 \\ \times.08 \\ \hline \end{array} \quad \begin{array}{r} 9.2 \\ \times\ .7 \\ \hline \end{array} \quad \begin{array}{r} .087 \\ \times\ .4 \\ \hline \end{array} \quad \begin{array}{r} 3.3 \\ \times.06 \\ \hline \end{array}$$

8.
$$\begin{array}{r} 41.8 \\ \times\ \ .7 \\ \hline \end{array} \quad \begin{array}{r} 3.90 \\ \times\ .08 \\ \hline \end{array} \quad \begin{array}{r} .516 \\ \times\ \ .5 \\ \hline \end{array} \quad \begin{array}{r} 73.8 \\ \times\ .06 \\ \hline \end{array} \quad \begin{array}{r} 3.47 \\ \times\ \ .4 \\ \hline \end{array}$$

9.
$$\begin{array}{r} 4.5 \\ \times 2.6 \\ \hline \end{array} \quad \begin{array}{r} .48 \\ \times 5.2 \\ \hline \end{array} \quad \begin{array}{r} .92 \\ \times 8.9 \\ \hline \end{array} \quad \begin{array}{r} 3.4 \\ \times.71 \\ \hline \end{array} \quad \begin{array}{r} .39 \\ \times.56 \\ \hline \end{array}$$

10.
$$\begin{array}{r} .27 \\ \times 1.8 \\ \hline \end{array} \quad \begin{array}{r} 5.6 \\ \times.39 \\ \hline \end{array} \quad \begin{array}{r} 4.9 \\ \times 5.4 \\ \hline \end{array} \quad \begin{array}{r} .83 \\ \times.17 \\ \hline \end{array} \quad \begin{array}{r} 9.2 \\ \times.66 \\ \hline \end{array}$$

11.
2.8	5.6	.72	.81	.94
×4.3	×.82	×5.7	×.69	×1.8

12.
30.5	7.40	61.8	.514	9.06
× .27	× 6.6	× 4.8	× .91	× 7.3

13.
45.21	3.748	206.9	.7488
× 5.6	× 73	× .28	× 4.9

14.
34.7	6.82	.925	8.43
× 209	×5.31	×78.8	×.409

15.
17.32	.0648	913.2	40.21
× .16	× 2.3	× .049	× 20.8

16.
.2789	3.845	5183.6	303.003
× .17	× 29.2	× .0016	× 56.8

Multiplying Decimals by 10, 100, and 1,000

There are shortcuts you can use when multiplying decimals by 10, 100 and 1,000. To multiply a decimal by 10, move the decimal point **one place to the right.**

EXAMPLE 1. $.26 \times 10$

Shortcut:

$.26 \times 10 = 2\,6 = 2.6$

Long way:

$$\begin{array}{r} .26 \\ \times\ 10 \\ \hline 2.6\cancel{0} \end{array}$$

Zeros at the end of a decimal are often dropped.

To multiply a decimal by 100, move the decimal point **two places to the right.**

EXAMPLE 2. 3.7×100

Shortcut:

$3.7 \times 100 = 3\,70 = 370$

Long way:

$$\begin{array}{r} 3.7 \\ \times 100 \\ \hline 370.\cancel{0} \end{array}$$

To multiply a decimal by 1,000, move the decimal point **three places to the right.**

EXAMPLE 3. $1.4 \times 1,000$

Shortcut:

$1.4 \times 1,000 = 1\,400 = 1,400$

Long way:

$$\begin{array}{r} 1.4 \\ \times 1,000 \\ \hline 1,400.\cancel{0} \end{array}$$

Multiply each of the following using the shortcut.

1. $.8 \times 10 =$ $.09 \times 10 =$ $3.64 \times 10 =$ $.721 \times 10 =$

2. $.03 \times 100 =$ $.275 \times 100 =$ $8.9 \times 100 =$ $.863 \times 100 =$

3. $.9 \times 1,000 =$ $2.36 \times 1,000 =$ $.475 \times 1,000 =$ $1.6 \times 1,000 =$

4. $.34 \times 10 =$ $1.24 \times 100 =$ $3.85 \times 1,000 =$ $.06 \times 1,000 =$

Multiplication of Decimals: Applying Your Skills

1. One inch is equal to 2.54 centimeters. How many centimeters are there in 12 inches?

2. Alberto makes $3.80 an hour. If he worked 7.5 hours on Monday, how much did he make that day?

3. One mile is equal to 1.6 kilometers. How many kilometers are there in 25 miles?

4. An airplane flew at an average speed of 386 miles per hour for .275 hours. How far did the plane fly?

5. One pound is equal to 0.45 kilograms. Find the weight in kilograms of a person who weighs 160 pounds.

6. If a cubic foot of water weighs 62.5 pounds, what is the weight of 16 cubic feet of water?

7. Jane makes $4.90 for an hour of over-time work. How much does she make for 2.3 hours of overtime work?

8. At $3.40 a meter, how much do 4.25 meters of lumber cost?

9. One yard is equal to 0.914 meters. How many meters are there in 75 yards?

10. It costs about .84¢ per hour to use a 100-watt lightbulb. How much does it cost to use it for 4.5 hours?

Dividing Decimals by Whole Numbers

To divide a decimal by a whole number, bring the point up in the answer directly above its position in the problem. Then divide as you would whole numbers.

EXAMPLE 1.

$$\begin{array}{r} 2.33 \\ 4\overline{)9.32} \\ 8 \\ \hline 1\,3 \\ 1\,2 \\ \hline 12 \\ 12 \\ \hline \end{array}$$

EXAMPLE 2.

$$\begin{array}{r} .037 \\ 6\overline{)\,.222} \\ 0 \\ \hline 22 \\ 18 \\ \hline 42 \\ 42 \\ \hline \end{array}$$

Divide the following.

1. $6\overline{)13.8}$ $9\overline{)70.2}$ $4\overline{)\,.384}$ $7\overline{)57.75}$

2. $8\overline{)\,.192}$ $3\overline{)148.8}$ $5\overline{)19.45}$ $4\overline{)2.524}$

3. $16\overline{)76.8}$ $21\overline{)7.56}$ $19\overline{)1.52}$ $38\overline{)216.6}$

4. $52\overline{)9.516}$ $43\overline{)1565.2}$ $77\overline{)464.31}$ $65\overline{)33.605}$

Dividing Decimals by Decimals

To divide by a decimal, you must change the problem to a problem in which you are dividing by a whole number.

EXAMPLE: $.03\overline{)4.374}$

Step 1. Move the point in the number outside the bracket (the divisor) to the right as far as it will go.

$.03\overline{)4.374}$

Step 2. Move the point in the number inside the bracket (the dividend) the same number of places that you moved the point in the divisor.

$.03\overline{)4.37\,4}$

Step 3. Bring the point up in the answer directly above its new position in the dividend and divide.

$1\,45.8$
$.03\overline{)4.37\,4}$

Divide the following.

1. $.8\overline{)7.68}$ $.7\overline{)26.6}$ $.6\overline{)0.42}$ $.4\overline{)2.76}$

2. $4.1\overline{)1.148}$ $9.2\overline{)128.8}$ $5.3\overline{)4.399}$ $3.2\overline{)0.288}$

3. $.06\overline{)0.882}$ $.28\overline{).5404}$ $.75\overline{)453.75}$ $.52\overline{)6.656}$

4. $8.7\overline{).522}$ $.19\overline{).3933}$ $3.6\overline{)145.44}$ $.64\overline{).11648}$

Sometimes you need to add zeros in your problem in order to have enough places to move the decimal point.

EXAMPLE: .08$\overline{)4.8}$

Step 1. Move the point in the divisor two places to the right.

.08$\overline{)4.8}$

Step 2. To move the point two places to the right in the dividend, add one zero.

.08$\overline{)4.80}$

Step 3. Bring the point up and divide.

$60.$
.08$\overline{)4.80}$

Divide the following.

5. .007$\overline{)5.32}$ .009$\overline{)4.32}$ .003$\overline{)8.1}$ .008$\overline{)77.12}$

6. .016$\overline{)212.8}$ .025$\overline{).1}$ .091$\overline{)65.52}$ .68$\overline{)57.8}$

7. .0024$\overline{).78}$ .0006$\overline{)42.84}$ .0018$\overline{)1.683}$ .0073$\overline{)15.184}$

8. .32$\overline{)156.8}$ .45$\overline{)1381.5}$ .008$\overline{)4.48}$ .06$\overline{)558.6}$

Dividing Whole Numbers by Decimals

When dividing a decimal into a whole number, put a point after the whole number and add zeros in order to move the point enough places. **Remember:** A whole number is understood to have a decimal point at its right.

EXAMPLE: $.007\overline{)35}$

Step 1. Move the point in the divisor three places to the right.

$.007\overline{)35}$

Step 2. Place a point to the right of the whole number and move it three places to the right, holding each place with a zero.

$.007\overline{)35.000}$

Step 3. Bring the point up and divide.

$$.007\overline{)35.000} = 5\,000.$$

Divide the following.

1. $.4\overline{)28}$ $.9\overline{)360}$ $.03\overline{)111}$ $.08\overline{)512}$

2. $.007\overline{)63}$ $.0005\overline{)4}$ $.06\overline{)234}$ $.3\overline{)1,884}$

3. $1.2\overline{)552}$ $.36\overline{)2,592}$ $4.27\overline{)2,135}$ $51.3\overline{)21,546}$

4. $.019\overline{)1,178}$ $.48\overline{)37,440}$ $5.6\overline{)33,040}$ $.039\overline{)3,237}$

Dividing Decimals by 10, 100, and 1,000

There are shortcuts you can use when dividing decimals by 10, 100, and 1,000. To divide a decimal by 10, move the decimal point **one place to the left.**

EXAMPLE 1. $7.2 \div 10$

Shortcut: $7.2 \div 10 = .7\,2 = .72$

Long way:

$$
\begin{array}{r}
.72 \\
10\overline{)7.20} \\
7\,0 \\
\hline
20 \\
20 \\
\hline
\end{array}
$$

To divide a decimal by 100, move the decimal point **two places to the left.**

EXAMPLE 2. $364 \div 100$

Shortcut: $364 \div 100 = 3\,64 = 3.64$

Long way:

$$
\begin{array}{r}
3.64 \\
100\overline{)364.00} \\
300 \\
\hline
64\,0 \\
60\,0 \\
\hline
4\,00 \\
4\,00 \\
\hline
\end{array}
$$

To divide a decimal by 1,000, move the decimal point **three places to the left.**

EXAMPLE 3. $25.3 \div 1,000$

Shortcut: $25.3 \div 1,000 = .025\,3 = .0253$

Long way:

$$
\begin{array}{r}
.0253 \\
1,000\overline{)25.3000} \\
00\,0 \\
\hline
25\,30 \\
20\,00 \\
\hline
5\,300 \\
5\,000 \\
\hline
3000 \\
3000 \\
\hline
\end{array}
$$

Divide each of the following using the shortcut.

1. $.9 \div 10 =$ $36 \div 10 =$ $27.3 \div 10 =$ $.04 \div 10 =$

2. $14.2 \div 100 =$ $1.3 \div 100 =$ $728 \div 100 =$ $.6 \div 100 =$

3. $37.5 \div 1,000 =$ $1.8 \div 1,000 =$ $2 \div 1,000 =$ $428 \div 1,000 =$

4. $13.45 \div 10 =$ $.32 \div 100 =$ $6,954 \div 1,000 =$ $15.8 \div 1,000 =$

Division of Decimals: Applying Your Skills

1. Together, the four members of the Rosa family weigh 508.8 pounds. What is the average weight of each of them?

2. Mark drove a total of 1,247.5 miles in five days. What was the average number of miles he drove each day?

3. If 3.5 yards of material cost $16.80, what is the price of one yard of the material?

4. Margie made $157.85 last week. If she makes $4.10 an hour, how many hours did she work last week?

5. If a plane flew 2,419.2 miles in 6.3 hours, what was its average speed in miles per hour?

6. If 2.3 pounds of beef cost $3.68, what is the price of one pound of beef?

7. There are 1.6 kilometers in a mile. How many miles are there in 36.8 kilometers?

8. There are 2.54 centimeters in one inch. How many inches are there in 45.72 centimeters?

Final Decimal Skills Inventory

1. Write fourteen ten-thousandths as a decimal.

2. Change .016 to a fraction and reduce.

3. Change $\frac{7}{16}$ to a decimal.

4. Which of the following is larger — .0013 or .02?

5. Rewrite the following list in order from smallest to largest: .03, .031, .1, .013

6. .056 + 2.3 + 19 =

7. 8 + .0573 + 1.64 =

8. In 1970, 19.2 thousand people lived in Smithtown, Ohio. By 1980, there were 3.9 thousand more people. How many people lived in Smithtown in 1980?

9. Janet bought 2.6 pounds of beef, 1.3 pounds of cheese, 2.45 pounds of chicken, and 5 pounds of sugar. What was the total weight she had to carry?

10. 12.3 − .097 =

11. 3 − .654 =

12. In polishing a piece of pipe that was 2 inches thick, .016 inch of metal was worn away. What was the thickness of the pipe when it was polished?

13. In 1974, the government counted 23.37 million people as being poor. In 1975, the government said there were 25.88 million people who were poor. How many more people did the government count as being poor in 1975 than in 1974?

14.   63.2
     ×.045

15.   91.12
     × 7.8

16. 20.8 × 100 =

17. One inch is equal to 2.54 centimeters. How many centimeters are there in 6.5 inches?

18. At $1.80 a meter, how much do 3.75 meters of wood cost?

19. 17) 69.53

20. .28) 2.016

21. 2.35) 14.1

22. .036$\overline{)9}$ **23.** 5.13 ÷ 1,000 =

24. A train traveled 144.9 miles in 4.2 hours. What was its average speed in miles per hour?

25. Manny made $24.70 for 6.5 hours of work. How much did he make each hour?

FINAL DECIMAL SKILLS INVENTORY CHART

A passing score is 20 problems correct. If you had less than 20 correct, review pages 49 through 69 before going on. Even if you had a passing score or better, any problem you missed should be corrected. Following is a list of the problems and the pages where each problem is covered.

Problem Number	Practice Page	Problem Number	Practice Page
1	51	14	59–61
2	52	15	59–61
3	53	16	62
4	54	17	63
5	54	18	63
6	55	19	64
7	55	20	65–66
8	56	21	65–66
9	56	22	67
10	57	23	68
11	57	24	69
12	58	25	69
13	58		

Percent Skills Inventory

Do all of the following problems that you can. There is no time limit. Work accurately, but do not use outside help.

1. Change .017 to a percent.

2. Change 4% to a decimal.

3. Change $\frac{5}{8}$ to a percent.

4. Change 12% to a common fraction.

5. Change $8\frac{1}{3}$% to a common fraction.

6. Find 15% of 80.

7. Find 4.6% of 250.

8. Find $37\frac{1}{2}$% of 96.

9. If the sales tax in a certain state is 8%, how much tax would you owe for a record that cost $4.50?

10. The Olsons spend $33\frac{1}{3}\%$ of their income on mortgage payments. If their income is $792 a month, what is their monthly mortgage payment?

11. 15 is what percent of 45?

12. 32 is what percent of 400?

13. On a loan of $500, Alfonso had to pay $55 in interest. The interest represents what percent of the loan?

14. At the meeting of a tenants' organization 36 people attended. The total membership of the organization is 54. What percent of the members attended the meeting?

15. 75% of what number is 60?

16. $62\frac{1}{2}\%$ of what number is 35?

17. The Morris family spends $160 a month for rent. If this represents 25% of their monthly income, what is their monthly income?

18. Chuck spent $.39 in tax for a shirt. If the tax rate in his state is 6%, what was the price of the shirt?

PERCENT SKILLS INVENTORY CHART

A passing score is 15 problems correct. If you had less than 15 correct, complete pages 76 through 95 before going on. Even if you had a passing score or better, any problem you missed should be corrected. Following is a list of the problems and the pages where each problem is covered.

Problem Number	Practice Page	Problem Number	Practice Page
1	76	10	83–84
2	77	11	85–86
3	78	12	85–86
4	79	13	87–88
5	79	14	87–88
6	81–82	15	89–90
7	81–82	16	89–90
8	81–82	17	91–92
9	83–84	18	91–92

What Are Percents?

Percent is a very common term in the everyday world. Commission, interest, mark-up, and tax rates are all written with percents. Discounts, raises, paycheck deductions, and credit card charges are all figured with percents.

Percent is another way to describe a part or fraction of something, but it is an even more special type of fraction. The *only* denominator (bottom number) it can have is 100. This denominator is not written; it is shown by a percent sign (%). For example, **49 parts out of 100** can be written as $\frac{49}{100}$ which is read as forty-nine hundredths, or as **.49** which is also read as forty-nine hundredths, or as **49%** which is read as forty-nine percent.

Changing Decimals to Percents

To change a decimal to a percent, move the decimal point two places to the **right** and write the percent sign (%). If the point moves to the end of the number, it is not necessary to write the point.

EXAMPLES:

Decimal		Move two places to the right		Percent
.35	=	.35	=	35%
.8	=	.80	=	80%
.04	=	.04	=	4%
$.12\frac{1}{2}$	=	$.12\frac{1}{2}$	=	$12\frac{1}{2}\%$
.0008	=	.0008	=	.08%

Change each decimal to a percent.

1. .32 =   .09 =   .6 =   .136 =

2. .005 = $.37\frac{1}{2}$ = $.08\frac{1}{3}$ =   .045 =

3. .0016 =   .0003 =   .025 = $.33\frac{1}{3}$ =

4. .125 =   .0375 =   .009 =   .8 =

Changing Percents to Decimals

To change a percent to a decimal, drop the percent sign and move the point two places to the left. Watch where zeros are necessary to move two places in the examples below.

EXAMPLES:

Percent		Move two places to the left		Decimal
6%	=	06	=	.06
30%	=	30	=	.3
150%	=	150	=	1.5
.9%	=	009	=	.009
$37\frac{1}{2}\%$	=	$37\frac{1}{2}$	=	$.37\frac{1}{2}$

Change each percent to a decimal.

1. 20% =   35% =   8% =   60% =

2. 3.5% =   .4% =   .03% =   21.6% =

3. $62\frac{1}{2}\%$ = $6\frac{2}{3}\%$ =   2.8% =   19% =

See how quickly you can fill in the following table. These are common decimals and percents, and you will save time later on if you know them automatically.

Percent	Decimal	Percent	Decimal	Percent	Decimal
50%		5%		37.5%	
25%		1%		62.5%	
75%		100%		87.5%	
20%		10%		12.5%	

Changing Fractions to Percents

There are two ways to change a fraction to a percent as shown below.

EXAMPLE: Change $\frac{3}{4}$ to a percent.

Method 1. Multiply the fraction by 100%.

$$\frac{3}{\cancel{4}_1} \times \frac{\overset{25}{\cancel{100\%}}}{1} = \frac{75\%}{1} = 75\%$$

Method 2. Divide the bottom number of the fraction into the top number and move the point two places to the right.

$$\frac{3}{4} = 4\overline{)3.00}^{.75} = 75\%$$

Change each fraction to a percent.

1. $\frac{2}{5} =$ $\frac{1}{4} =$ $\frac{1}{3} =$ $\frac{3}{8} =$

2. $\frac{6}{25} =$ $\frac{2}{3} =$ $\frac{5}{6} =$ $\frac{1}{8} =$

3. $\frac{9}{10} =$ $\frac{7}{8} =$ $\frac{11}{20} =$ $\frac{5}{12} =$

4. $\frac{1}{6} =$ $\frac{4}{5} =$ $\frac{7}{10} =$ $\frac{1}{12} =$

5. $\frac{5}{8} =$ $\frac{4}{9} =$ $\frac{3}{7} =$ $\frac{9}{20} =$

6. $\frac{4}{25} =$ $\frac{3}{10} =$ $\frac{3}{5} =$ $\frac{9}{50} =$

Changing Percents to Fractions

To change a percent to a fraction, write the percent as a fraction with 100 as the bottom number and reduce.

EXAMPLE 1. Change 85% to a fraction.

Step 1. Write the percent as a fraction with 100 as the bottom number.

$$\frac{85}{100}$$

Step 2. Reduce.

$$\frac{85 \div 5}{100 \div 5} = \frac{17}{20}$$

EXAMPLE 2. Change $8\frac{1}{3}\%$ to a fraction.

Step 1. Write the percent as a fraction with 100 as the bottom number.

$$\frac{8\frac{1}{3}}{100}$$

Step 2. You can rewrite this fraction as a division problem.

$$8\frac{1}{3} \div 100$$

Step 3. Change the mixed number to an improper fraction.

$$\frac{25}{3} \div \frac{100}{1}$$

Step 4. Invert the divisor and multiply, canceling where possible.

$$\overset{1}{\cancel{\frac{25}{3}}} \times \frac{1}{\underset{4}{\cancel{100}}} = \frac{1}{12}$$

Change each percent to a common fraction.

1. $35\% =$ 　　　　 $20\% =$ 　　　　 $12\frac{1}{2}\% =$ 　　　　 $6\% =$

2. $16\frac{2}{3}\% =$ 　　　　 $1\% =$ 　　　　 $90\% =$ 　　　　 $37\frac{1}{2}\% =$

3. $12\% =$ 　　　　 $99\% =$ 　　　　 $66\frac{2}{3}\% =$ 　　　　 $4\frac{1}{2}\% =$

4. $80\% =$ 　　　　 $33\frac{1}{3}\% =$ 　　　　 $4\% =$ 　　　　 $8\frac{1}{3}\% =$

Common Percents and Their Values as Proper Fractions

After you have filled in the table on this page, check your answers. Then memorize the table. These are the most common fractions and percents. You will save time later on if you know what each of them is equal to.

Fraction	Percent		Fraction	Percent
$\frac{1}{2}$ =	50%		$\frac{1}{8}$ =	
$\frac{1}{4}$ =			$\frac{3}{8}$ =	
$\frac{3}{4}$ =			$\frac{5}{8}$ =	
			$\frac{7}{8}$ =	
$\frac{1}{5}$ =			$\frac{1}{10}$ =	
$\frac{2}{5}$ =			$\frac{3}{10}$ =	
$\frac{3}{5}$ =			$\frac{7}{10}$ =	
$\frac{4}{5}$ =			$\frac{9}{10}$ =	
$\frac{1}{3}$ =			$\frac{1}{6}$ =	
$\frac{2}{3}$ =			$\frac{5}{6}$ =	

Do your work here.

Finding a Percent of a Number

To find a percent of a number, change the percent to a decimal or to a fraction and multiply.

EXAMPLE: Find 25% of 80.

Method 1.

    Step 1. Change the percent to a decimal. $25\% = .25$

    Step 2. Multiply.

$$\begin{array}{r} 80 \\ \times\ .25 \\ \hline 4\ 00 \\ 16\ 0 \\ \hline 20.00 \end{array}$$

Method 2.

    Step 1. Change the percent to a fraction. $25\% = \dfrac{1}{4}$

    Step 2. Multiply.

$$\frac{1}{\overset{}{\underset{1}{4}}} \times \frac{\overset{20}{80}}{1} = \frac{20}{1} = 20$$

Use the method that you find easier to work the following.

1.   5% of 120 =   7% of 965 =   10% of 780 =

2.   20% of 36 =   15% of 50 =   40% of 60 =

3.   75% of 680 =   80% of 500 =   50% of 418 =

4.   35% of 480 =   65% of 620 =   85% of 940 =

5.   2.6% of 390 =   .8% of 56 =   1.8% of 753 =

If you want to multiply by a complex percent like $16\frac{2}{3}\%$, it is easiest to change the percent to the fraction that it is equal to (from the table on page 80) and then multiply.

EXAMPLE 1.   Find $16\frac{2}{3}\%$ of 42.

Step 1.   Change the complex percent to a fraction.

$$16\frac{2}{3}\% = \frac{1}{6}$$

Step 2.   Multiply.

$$\frac{1}{\cancel{6}_1} \times \frac{\cancel{42}^7}{1} = \frac{7}{1} = 7$$

If you do not know the fractional value of a complex percent, multiply by the improper fraction form of the percent, and put the other number over 100.

EXAMPLE 2.   Find $6\frac{2}{3}\%$ of 45.

Step 2.   Change the numbers of the percent to an improper fraction.

$$6\frac{2}{3} = \frac{20}{3}$$

Step 2.   Put the other number over 100 and multiply.

$$\frac{\cancel{20}^1}{\cancel{3}_1} \times \frac{\cancel{45}^{15}}{\cancel{100}_5} = \frac{15}{5} = 3$$

Find the following.

1. $33\frac{1}{3}\%$ of 75 = $12\frac{1}{2}\%$ of 96 = $16\frac{2}{3}\%$ of 84 =

2. $37\frac{1}{2}\%$ of 720 = $83\frac{1}{3}\%$ of 630 = $1\frac{1}{2}\%$ of 200 =

3. $5\frac{1}{4}\%$ of 400 = $66\frac{2}{3}\%$ of 90 = $87\frac{1}{2}\%$ of 200 =

4. $8\frac{1}{3}\%$ of 36 = $6\frac{1}{4}\%$ of 60 = $62\frac{1}{2}\%$ of 176 =

Finding a Percent of a Number: Applying Your Skills

When you are finding a percent of a number, you are always looking for a **part** of that number. Be sure to label your answers with the same units as given in the problem.

ANSWERS

_____ 1. During a period of 30 working days, Deborah was late 80% of the time. How many days was she late?

_____ 2. For selling their house, Mr. and Mrs. Martin had to pay the real estate agent a commission of 5%. If their house sold for $35,000, how much commission did the real estate agent receive?

_____ 3. A coat originally selling for $48 was on sale at "20% off." How much can someone save by buying the coat on sale?

_____ 4. If the sales tax in a certain state is 6%, how much tax would you owe for a scarf that cost $2.50?

_____ 5. How many answers did Linda leave blank if she left out 15% of the questions on a job application with a total of 40 questions?

_____ 6. 540 workers belong to the local of an electricians' union. If 65% of them went to the last meeting, how many members went to the meeting?

_____ 7 Paul makes a gross salary of $160 each week. If 18% of his salary is withheld for taxes and social security, how much is withheld from his weekly check?

_____ 8. Doreen's night class met 30 times. If she was absent from 20% of the classes, how many times was she absent?

_____ 9. Elizabeth makes $118 a week. If she gets an 8% raise, how much more will she make each week?

_____ 10. A portable television selling for $160 was on sale for $12\frac{1}{2}\%$ less. How much can be saved by buying the television on sale?

_____ 11. Ruby lost 6.5% of her weight. How many pounds did she lose if she used to weigh 150 pounds?

_____ 12. Ellen, who made $15,000 a year, got a raise of 8.5%. How much more will she make in a year?

_____ 13. Kate gets a 15% discount on the price of anything in the store where she works. How much will she save on a dress that is priced at $28?

_____ 14. The Smiths spend $33\frac{1}{3}\%$ of their income for food. If they make $168 a week, how much do they spend each week for food?

_____ 15. Calvin made a down payment of $12\frac{1}{2}\%$ on a car which cost $2,400. How much was the down payment?

_____ 16. Joe was supposed to work 250 days last year. He was absent 4% of the time because he was sick. How many days did he miss?

Finding What Percent One Number Is of Another

To find what percent one number is of another, make a fraction by putting the part (usually the smaller number) over the whole. Reduce the fraction and change it to a percent.

EXAMPLE: 9 is what % of 45?

Step 1.   Put the part over the whole and reduce. $\frac{9}{45} = \frac{1}{5}$

Step 2.   Divide the bottom number into the top (see page 78).

$$\overset{.20}{5\overline{)1.00}}$$

Step 3.   Change your decimal answer to a percent by moving two places to the right and adding a percent (%) sign. $.20 = 20\%$

Find the following.

1. 8 is what % of 16?   15 is what % of 60?   27 is what % of 81?

2. 9 is what % of 90?   12 is what % of 72?   40 is what % of 320?

3. 16 is what % of 20?   14 is what % of 35?   32 is what % of 48?

4. 75 is what % of 90?   33 is what % of 44?   56 is what % of 64?

5. 35 is what % of 40?   45 is what % of 50?   160 is what % of 200?

6. 14 is what % of 200?   24 is what % of 400?   45 is what % of 500?

7. 7 is what % of 20?   19 is what % of 50?   8 is what % of 25?

8. 21 is what % of 36?   19 is what % of 19?   12 is what % of 27?

9. 60 is what % of 72?   24 is what % of 56?   126 is what % of 140?

10. 4 is what % of 200?   18 is what % of 300?   12 is what % of 144?

11. 15 is what % of 75?   84 is what % of 105?   27 is what % of 120?

12. 12 is what % of 72?   2,600 is what % of 10,000?   792 is what % of 200,000?

Finding What Percent One Number Is of Another: Applying Your Skills

When you are finding what percent one number is of another, always be careful to put the **part** over the **whole**.

ANSWERS

_____ 1. There are 24 tenants in a community block association. If only 18 of them came one night, what percent of the association was there?

_____ 2. Joaquin makes $150 a week. If he gets a raise of $12 each week, what percent of his original salary was his raise?

_____ 3. Floria wants to buy a coat that costs $48. If she has already saved $30, what percent of the price of the coat has she saved?

_____ 4. Matthew got 36 problems right out of a total of 40 problems on a test. What percent of the problems did he get right?

_____ 5. On a loan of $350, Manny had to pay $28 of interest. The interest represents what percent of Manny's loan?

_____ 6. Carl weighed 180 pounds. After two months of dieting and exercising, he lost 9 pounds. What percent of his weight did he lose?

_____ 7. Margaret is reading a book that is 288 pages long. If she has already read 96 pages, what percent of the book has she read?

_____ **8.** The Johnson family spends $135 a month on rent. If their monthly income is $540, what percent of their income goes to rent?

_____ **9.** In an office of 20 employees, there are 11 men. What percent of the office is made up of men?

_____ **10.** Carlos owed $2,000 to the bank. If he has paid back $1,250, what percent of the loan has he paid back?

_____ **11.** The price of a dozen large eggs went up 7¢ from 84¢. The increase represents what percent of the original price?

_____ **12.** The Rivera family makes $180 a week. If they spend $45 each week for food, what percent of their income goes for food?

_____ **13.** George bought a new car last year for $3,520. This year it is worth $440 less. The decrease in value represents what percent of the original price?

_____ **14.** Last year the population of Middleville was 16,000 people. In one year the population has increased by 800 people. The increase represents what percent of last year's population?

_____ **15.** Rachel has been sick 8 days out of the past 48. What percent of the time has she been sick?

_____ **16.** An inch represents what percent of a foot?

Finding a Number When a Percent of It Is Given

If a percent of a number is given and you are looking for the whole number, change the percent into either a fraction or a decimal and divide it into the number you have.

EXAMPLE: 20% of what number is 16?

Method 1.

Step 1. Change the percent to a fraction.

$$20\% = \frac{1}{5}$$

Step 2. Divide the fraction into the number you have.

$$16 \div \frac{1}{5} = \frac{16}{1} \times \frac{5}{1} = \frac{80}{1} = 80$$

Method 2.

Step 1. Change the percent to a decimal.

$$20\% = .2$$

Step 2. Divide the decimal into the number you have.

$$.2\,\overline{)\,16.0\,}^{\;80.}$$

Use the method you find easier to work the following.

1. 25% of what number is 8?   50% of what number is 45?

2. 75% of what number is 48?   60% of what number is 75?

3. 40% of what number is 60?   15% of what number is 12?

4. 10% of what number is 6.3?   35% of what number is 8.4?

89

5. $33\frac{1}{3}\%$ of what number is 25? $12\frac{1}{2}\%$ of what number is 160?

6. $16\frac{2}{3}\%$ of what number is 300?   8.5% of what number is 119?

7. $37\frac{1}{2}\%$ of what number is 24? $83\frac{1}{3}\%$ of what number is 150?

8. $66\frac{2}{3}\%$ of what number is 320? $87\frac{1}{2}\%$ of what number is 210?

9. $62\frac{1}{2}\%$ of what number is 120?   12.6% of what number is 189?

10. 80% of what number is 244? $33\frac{1}{3}\%$ of what number is 350?

11. $83\frac{1}{3}\%$ of what number is 365?   20% of what number is 175?

Finding a Number When a Percent of It Is Given: Applying Your Skills

In problems of this type, always divide the percent (in either decimal or fraction form) into the number that you have. Your answer will usually be larger than the original number.

ANSWERS

_____ 1. Morris has saved $80 toward buying an air conditioner. If the $80 is 50% of the total price, what is the price of the air conditioner?

_____ 2. Lois got 27 problems right on a math test. If this is 90% of the test, how many problems were on the test?

_____ 3. Pat made a down payment of $500 on a used car. The $500 is 20% of the price of the car. Find the price of the car.

_____ 4. The Rigby family spends an average of $56 a week for food. If this is $33\frac{1}{3}$% of their weekly budget, what is the amount of their weekly budget?

_____ 5. Phil had to pay $4.80 in tax for a new suit. If the sales tax in Phil's state is 8%, what was the price of the suit?

_____ 6. 78 members came to a club meeting. If they are 65% of the total membership, how many members are in the club?

_____ 7. Mavis made $85 in commissions one week for selling shoes. If her commission rate is 5% of the total amount that she sells, what was the value of the shoes she sold that week?

_____ 8. When the Greens bought their house, they made a down payment of $5,100, which was 15% of the total cost of the house. Find the cost of the house.

_____ 9. After dieting for two months, Sonia lost 18 pounds, which was 8% of her original weight. What was her original weight?

_____ 10. 3,000 people came to hear a presidential candidate speak in Central City. If they are $16\frac{2}{3}\%$ of the people who live in Central City, how many people live there?

_____ 11. David has driven 2,550 miles on his way from New York to San Francisco. If this is 85% of the total distance, what is the distance from New York to San Francisco?

_____ 12. Tom had to pay $240 in interest on a loan. If $240 is 12% of the total loan, how much was the loan?

_____ 13. Mr. Moore pays 52¢ for a dozen small eggs. 52¢ is 65% of the amount he charges his customers. How much do his customers pay for a dozen small eggs?

_____ 14. Walter's weekly take-home pay is $151.32, which is 78% of his gross pay (before deductions). What is Walter's weekly gross pay?

Final Percent Skills Inventory

1. Change .123 to a percent.

2. Change 5.6% to a decimal.

3. Change $\frac{7}{12}$ to a percent.

4. Change 84% to a fraction.

5. Change $31\frac{1}{4}$% to a fraction.

6. Find 9% of 270.

7. Find 11.3% of 460.

8. Find $8\frac{1}{3}$% of 72.

9. By working for a certain factory, Bill gets a 15% discount on washers. How much can he save on a washer that sells for $165?

10. There are 24,000 voters in Millville. If $37\frac{1}{2}\%$ of them voted in the last election, how many people voted?

11. 45 is what percent of 72?

12. 17 is what percent of 200?

13. The Millers want to buy a house for which the down payment is $5,000. If they have already saved $3,750, what percent of the down payment have they saved?

14. Of the 32 employees in an office, 20 of them are women. What percent of the office staff is women?

15. 40% of what number is 80?

16. $12\frac{1}{2}\%$ of what number is 48?

17. Alfredo has driven 273 miles which is 65% of the distance he has to drive. How far does he have to drive?

18. Norma typed 28 letters in three days. If this is $87\frac{1}{2}\%$ of her work for the week, how many letters does she have to type all together?

FINAL PERCENT SKILLS INVENTORY CHART

A passing score is 15 problems correct. If you had less than 15 correct, review pages 76 through 92 before going on. Even if you had a passing score or better, any problem you missed should be corrected. Following is a list of the problems and the pages where each problem is covered.

Problem Number	Practice Page	Problem Number	Practice Page
1	76	10	83–84
2	77	11	85–86
3	78	12	85–86
4	79	13	87–88
5	79	14	87–88
6	81–82	15	89–90
7	81–82	16	89–90
8	81–82	17	91–92
9	83–84	18	91–92

Building Number Power: Review Test

The purpose of this test is for you to see how well you have mastered all of the skills you practiced in this section of the book. Take your time and work each problem carefully.

---

1. Reduce $\frac{16}{36}$

2.
$$5\frac{7}{8}$$
$$4\frac{2}{3}$$
$$+9\frac{1}{2}$$

3. A table $46\frac{1}{2}$ inches long can be made longer with an extra piece that is $12\frac{2}{3}$ inches long. How long is the table with the extra piece?

4.
$$23\frac{2}{9}$$
$$-18\frac{3}{4}$$

5. How much longer is a pole $75\frac{1}{4}$ inches long than one that is $68\frac{5}{16}$ inches long?

6. $\frac{5}{6} \times \frac{8}{15} \times \frac{3}{20} =$

7. $4\frac{1}{2} \times 2\frac{2}{5} \times 3\frac{3}{4} =$

8. If a plumber charges $9.00 an hour for his work, how much would he charge for a job that took him $1\frac{3}{4}$ hours to finish?

9. $\frac{3}{5} \div 15 =$

10. $4\frac{2}{3} \div 1\frac{1}{9} =$

11. How many $\frac{3}{4}$-pound bags of sugar can be filled from 36 pounds of sugar?

12. Write fifty and two hundred eight thousandths as a mixed decimal.

13. Rewrite the following list in order from smallest to largest: .021, .12, .2, .02

14. $293.08 + 14 + 2.719 =$

15. Mary, who weighed 119.3 pounds in September of 1987, weighed 8.8 pounds more in February of 1988. What was her weight in February?

16. 40 − .387 =

17. The Rodriguez family plans to drive 250 miles to visit their cousins. If they have already driven 113.8 miles, how much farther do they have to drive?

18.   2.048
   × 1.9

19.   67.4
   × .08

20. Find the cost of 4.8 kilograms of tomatoes at $.75 a kilogram?

21. 4.9$)\overline{1.274}$

22. .018$)\overline{.3474}$

23. There are 1.6 kilometers in a mile. How many miles are there in 24 kilometers?

24. Change 7.9% to a decimal.

25. Change $\frac{9}{16}$ to a percent.

26. Change $16\frac{2}{3}\%$ to a fraction.

27. Find 18% of 304.

28. Find $8\frac{1}{3}\%$ of 144.

29. Mike sells sporting goods on an 8% commission. If he sold $1,384 worth of sporting goods in one week, what was his commission for the week?

30. 75 is what percent of 90?

31. Sandy owed the bank $1,200. She has already paid back $750. What percent of the money that she owed has she paid back to the bank?

32. 35% of what number is 56?

33. A community organization has raised $200,000 to build a youth center. The amount is 80% of the total that they need. How much do they need to build the youth center?

BUILDING NUMBER POWER: SKILLS CHART

A passing score is 27 problems correct. If you had less than 27 correct, review the first three sections of this book before going on to USING NUMBER POWER. Even if you had a passing score or better, any problem you missed should be corrected. Following is a list of the problems and the pages where each problem is covered. Review the pages that cover the problems you missed.

Problem Number	Practice Page	Problem Number	Practice Page
1	8–9	18	59–61
2	16–19	19	59–61
3	20	20	63
4	22–26	21	65–66
5	27	22	65–66
6	28–30	23	69
7	32	24	77
8	33	25	78
9	38–39	26	79
10	40–41	27	81
11	42	28	82
12	51	29	83–84
13	54	30	85–86
14	55	31	87–88
15	56	32	89–90
16	57	33	91–92
17	58		

Using
Number Power

Reading a Ruler

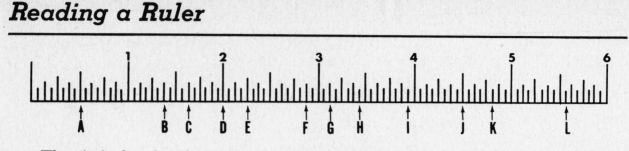

The six-inch ruler shown on this page is divided into inches (at the numbers), half inches (every eight lines), quarter inches (every four lines), eighth inches (every other line), and sixteenth inches (from one line to the next).

Remember that rulers, like books, are read from left to right. The arrow at A is $\frac{1}{2}$ inch from the left end of the ruler. The arrow at B is $1\frac{3}{8}$ inches from the left end.

1. Tell in inches how far from the left end each of the following points is.

        C_____   H_____

        D_____   I_____

        E_____   J_____

        F_____   K_____

        G_____   L_____

2. How much farther is point D from the left end of the ruler than point C?

3. How much farther is point H from the left end of the ruler than point F?

4. How much farther is point L from the left end of the ruler than point J?

Reading a Metric Ruler

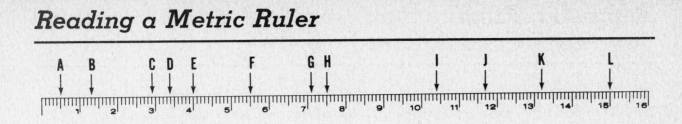

The sixteen-centimeter ruler shown on this page is divided into centimeters (at the numbers) and millimeters at each line. There are ten millimeters in a centimeter. Every fifth millimeter is a longer mark on the ruler.

The arrow at A is 5 millimeters or .5 centimeters from the left end of the ruler. The arrow at B is 1.3 centimeters from the left end.

---

1. Tell in centimeters how far from the left end each of the following points is.

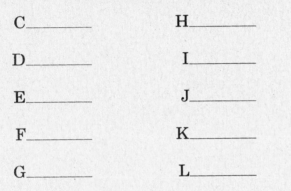

           C_____   H_____

           D_____   I_____

           E_____   J_____

           F_____   K_____

           G_____   L_____

2. How much farther is point E from the left end of the ruler than point C?

3. How much farther is point G from the left end of the ruler than point F?

4. How much farther is point K from the left end of the ruler than point J?

Perimeter: Finding the Distance Around Something (Rectangles and Squares)

When you measure the distance around something, you are finding its **perimeter.** You can find this distance (perimeter) by measuring each side and adding all the measurements together. If you want to find the perimeter of something with two pairs of equal sides (a **rectangle**), you can double the length (the long sides), double the width (the short sides), and add these two answers. This rule can be written in a short form called a formula: $P = 2l + 2w.$ **P** stands for perimeter; **l** stands for length; and **w** stands for width.

EXAMPLE: How much tape do you need to go around the edges of a box with a length of $4\frac{3}{4}$ inches and a width of $2\frac{1}{2}$ inches.

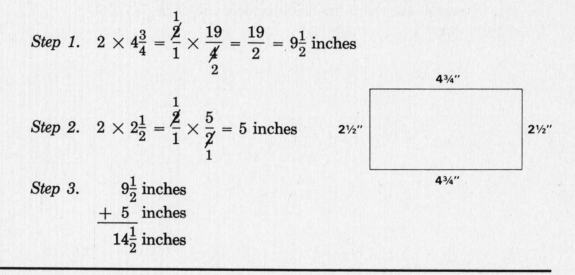

Step 1. $2 \times 4\frac{3}{4} = \frac{\cancel{2}^{1}}{1} \times \frac{19}{\cancel{4}_{2}} = \frac{19}{2} = 9\frac{1}{2}$ inches

Step 2. $2 \times 2\frac{1}{2} = \frac{\cancel{2}^{1}}{1} \times \frac{5}{\cancel{2}_{1}} = 5$ inches

Step 3.
$$
\begin{array}{r}
9\frac{1}{2} \text{ inches} \\
+ \ 5 \ \text{ inches} \\
\hline
14\frac{1}{2} \text{ inches}
\end{array}
$$

---

1. How much tape do you need to go around the edges of a box $7\frac{2}{3}$ inches long and 3 inches wide?

2. How much weather stripping will you need to go around a window that is $6\frac{7}{8}$ feet long and $2\frac{1}{4}$ feet wide.

3. How many inches of picture frame molding are needed to go around a picture that is 11 inches long and $8\frac{1}{2}$ inches wide?

4. Mr. King's garden is 6.2 meters long and 5.8 meters wide. How many meters of fencing are needed to enclose the garden?

5. If fencing costs $4.50 per meter, how much would the fencing for the garden in problem 4 cost?

6. Find the perimeter of a tabletop that is $62\frac{1}{2}$ inches long and $30\frac{1}{4}$ inches wide.

7. What is the perimeter of a photograph that is 10.3 centimeters long and 4.6 centimeters wide?

8. How much fencing would be needed to enclose a garden that measures $6\frac{1}{4}$ yards on every side?

9. Find the perimeter of a snapshot that measures $2\frac{1}{2}$ inches on each side.

10. What is the perimeter of a piece of tile that measures 1.75 meters on every side?

Area: Finding the Amount of Space Taken Up by Something (Rectangles and Squares)

When you buy carpeting, you have to know how much space you want to cover with it. To figure out the answer to problems like this, you need to find **area**. Area is the amount of space or surface of a flat figure like a floor. Area is measured in square inches, square feet, square yards, etc. To find the area of a rectangle, you multiply the length by the width. As a formula the rule is: **A = lw. A** stands for the area; **l** stands for the length; and **w** stands for the width. Notice that **l** and **w** written next to each other means that they should be multiplied together.

EXAMPLE: How much shelf paper do you need to cover a shelf that is $4\frac{1}{2}$ feet long and $1\frac{1}{4}$ feet wide.

$$4\frac{1}{2} \times 1\frac{1}{4} = \frac{9}{2} \times \frac{5}{4} = \frac{45}{8} = 5\frac{5}{8} \text{ square feet}$$

Be sure to label each answer in square units.

---

1. How much glass do you need to cover a coffee table that is $5\frac{1}{3}$ feet long and $3\frac{3}{4}$ feet wide?

2. How much space is covered by a tile that is 6.2 centimeters long and 2.8 centimeters wide?

3. How many square inches of plastic are needed to cover a photograph that is $7\frac{1}{2}$ inches long and 4 inches wide?

4. Find the area of a floor that is 7.2 meters long and 5.5 meters wide.

5. If one square meter of carpet costs $6.40, how much would the carpet cost to cover the floor in problem 4?

6. How many square yards of carpeting are needed to cover the floor of a room that is $3\frac{1}{3}$ yards long and $2\frac{1}{4}$ yards wide?

7. How much space can you cover with a piece of linoleum that measures 3.6 meters on each side?

8. What is the area of a tablecloth needed to cover a tabletop that measures $2\frac{1}{3}$ feet on each side?

9. How much glass is needed to cover the front of a photograph that measures $3\frac{1}{2}$ inches on each side?

Volume: Finding the Amount of Space Inside Something (Rectangular Shapes)

To find out how much space is **inside** things like boxes, suitcases, and rooms, you have to find volume. Volume is measured in cubic inches, cubic feet, cubic yards, etc. To find the volume of things shaped like a rectangle, such as a shoebox, you multiply the length by the width by the height of the object. As a formula the rule is: $V = lwh$. V stands for the volume; l stands for the length; w stands for the width; and h stands for the height.

EXAMPLE: Find the volume of a cardboard box that is 8 inches long, $4\frac{1}{2}$ inches wide, and $3\frac{1}{3}$ inches high.

$$8 \times 4\frac{1}{2} \times 3\frac{1}{3} = \frac{\overset{4}{\cancel{8}}}{1} \times \frac{\overset{3}{\cancel{9}}}{\underset{1}{\cancel{2}}} \times \frac{10}{\underset{1}{\cancel{3}}} = 120 \text{ cubic inches}$$

Be sure to label each volume answer in cubic units.

---

1. Find the volume of a box that is 3 feet long, $2\frac{1}{2}$ feet wide, and $1\frac{1}{3}$ feet high.

2. Find the volume of a water tank that is 10 feet long, 8.5 feet wide, and 6.25 feet deep.

3. How many cubic feet of air are there in a room that is 12 feet long, 9 feet wide, and $8\frac{1}{2}$ feet high?

4. How much space is there inside a packing crate that is 16 inches long, $12\frac{1}{2}$ inches wide, and $7\frac{1}{2}$ inches high?

5. What is the volume of a matchbox that is 4 inches long, $2\frac{1}{2}$ inches wide, and $\frac{3}{4}$ inch high?

6. How many of the matchboxes in problem 5 fit inside the packing crate in problem 4?

7. What is the volume of a hole for a building foundation that is 20.4 meters long, 16.5 meters wide, and 3.8 meters deep?

8. How much concrete is in a sidewalk that is 63 feet long, 4 feet wide, and $\frac{1}{3}$ foot deep?

9. How much packing space is there in a trunk that is $4\frac{1}{3}$ feet long, $2\frac{1}{4}$ feet wide, and $2\frac{2}{3}$ feet high?

Circumference: Finding the Distance Around a Circle

The distance across a circle is called the **diameter,** which can be shown by a straight line that goes through the center of a circle. The distance around a circle is called the **circumference.** To find the circumference of a circle, multiply the diameter of the circle by π (pi), a special number that has no exact value but is close to $3\frac{1}{7}$ written as a fraction, or 3.14 written as a decimal.

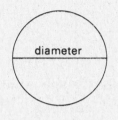

As a formula, the distance around a circle is:

$C = \pi d$. C is the circumference; π is $3\frac{1}{7}$ or 3.14; and **d** is the diameter. Notice that π and **d** written next to each other means that they should be multiplied together.

EXAMPLE: Find the circumference of a circular chair seat with a diameter of 21 inches.

$$3\frac{1}{7} \times 21 = \frac{22}{\underset{1}{7}} \times \frac{\overset{3}{\cancel{21}}}{1} = 66 \text{ inches}$$

Use $\pi = 3\frac{1}{7}$ to work these problems.

1. Find the circumference of a circle with a diameter of 56 yards.

2. Mr. Casa wants to put a fence around a circular space in his back yard for his children to play in. If the space has a diameter of 35 feet, how many feet of fencing does he need?

3. If the fencing for problem 2 costs $.85 a foot, how much will the fencing cost to enclose the space in Mr. Casa's back yard?

110

4. What is the circumference of a hubcap with a diameter of 10 inches?

5. Charles built a round table with a diameter of $3\frac{1}{2}$ feet. He wants to put metal stripping around the edge of the table. How many feet of stripping must he buy?

6. If the stripping for problem 5 costs 29¢ a foot, how much will Charles pay for the stripping to go around his table?

7. How many inches of wood are needed to make a frame for a circular mirror with a diameter of 28 inches?

8. How much fringe would be needed to trim a lampshade that has a diameter of 14 inches?

9. How much fencing is needed to go around a circular pool that has a diameter of 30 feet?

Area: Finding the Amount of Space Taken Up by a Circle

You already learned that the distance across a circle is called the diameter. The distance halfway across a circle is called the **radius**. The radius is $\frac{1}{2}$ the diameter. To find the area of a circle, multiply the radius by itself and then multiply this answer by π. The formula for finding the area of a circle is $A = \pi r^2$. **A** is the area; π is $3\frac{1}{7}$ or 3.14; and r^2 is the radius multiplied by itself.

EXAMPLE: Find the area of a circular hole in the floor with a radius of 2 inches.

$$\frac{22}{7} \times \frac{2}{1} \times \frac{2}{1} = \frac{88}{7} = 12\frac{4}{7} \text{ square inches}$$

Notice that the answer is in square units.

Use $\pi = 3\frac{1}{7}$ to work these problems. Be sure to label area answers in square units.

1. Find the area of a circle with a radius of 7 feet.

2. If the radius of a circular pool is 10 feet, what is the area of the bottom of the pool?

3. Mrs. Twyman wants to cover the top of a round table with square tiles that are each 1-inch square. How many tiles will she need if the radius of her table is 14 inches?

4. If the tiles for the table in problem 3 cost 2¢ each, how much will Mrs. Twyman pay for the tiles to cover her table?

5. How large a bedspread is needed to cover a round bed with a radius of 2.1 meters?

6. How many square yards of dancing space are there on a round dance floor with a radius of $3\frac{1}{2}$ yards?

7. If flooring costs $4.50 per square yard, how much will the flooring cost to cover the dance floor?

8. Find the amount of space inside a circular garden with a radius of one foot.

9. A circular hole in Mr. Henry's wall has a radius of one-half foot. How much material does he need to cover it?

Changing a Recipe

Use the following list of things needed for making a cake to answer the questions on this page.

1 cup of flour	$1\frac{1}{3}$ cups of egg whites
$1\frac{1}{2}$ cups of sugar	$1\frac{2}{3}$ teaspoons of cream of tartar
$\frac{1}{4}$ teaspoon of salt	$1\frac{1}{4}$ teaspoons of vanilla

1. Fill in the amount of each thing you would need to make two cakes.

   _____flour   _____egg whites

   _____sugar   _____cream of tartar

   _____salt   _____vanilla

2. Fill in the amount of each thing you would need to make a smaller cake that is one-half the size of the cake in the recipe.

   _____flour   _____egg whites

   _____sugar   _____cream of tartar

   _____salt   _____vanilla

3. Fill in the amount of each thing you would need to make 5 cakes.

   _____flour   _____egg whites

   _____sugar   _____cream of tartar

   _____salt   _____vanilla

How Much Do You Pay?
Rounding Off Money to the Nearest Cent

There are many everyday math problems dealing with money that end up with more than two decimal places in the answer. Since our money has only two decimal places (dimes are in the tenths place, and pennies are in the hundredths place), you often have to **round off** an answer to the nearest cent.

Use these two rules to round off an answer to the nearest cent:

If the thousandths (third) decimal place is **4 or less, leave** the answer with the hundredths place as it is.

If the thousandths place is **5 or more, raise** the answer to the next cent.

EXAMPLE 1.   If the sales tax rate in a certain state is 6%, how much tax would you owe on a hat that costs $4.59?

Step 1.   6% = .06 $4.59
                        × .06

                      $.2754

Step 2.   Since the thousandths place has a 5, raise the answer to $.28.

EXAMPLE 2.   Find the cost of 9 gallons of gasoline at 57.6¢ per gallon.

Step 1.   57.6¢ = $.576 $.576
                           ×   9

                         $5.184

Step 2.   Since the thousandths place has a 4, leave the answer as $5.18.

EXAMPLE 3.   If the sales tax rate in a certain state is 5%, how much tax would you owe on a record that costs $7.95?

Step 1.   5% = .05 $7.95
                        × .05

                      $.3975

Step 2.   Since the thousandths place has a 7, raise the answer to the next cent: $.40.

1. The sales tax rate in a certain state is 6%. How much tax would you owe on a sweater that costs $12.49?

2. Find the cost of 8 gallons of gasoline at 59.2¢ per gallon.

3. Find the cost of 12 gallons of gasoline at 61.6¢ per gallon.

4. Shirley bought two shirts at $6.59 each. How much sales tax does she owe on the two shirts if the tax rate in her state is 7%?

5. What was the total amount that Shirley, in problem 4, had to pay for the shirts including tax?

6. Sam bought a portable television for $69.58. How much sales tax did he owe if the tax rate in his state is 8%?

7. What was the total amount that Sam, in problem 6, had to pay for the television including tax?

8. Find the cost of 7 gallons of gasoline at 58.5¢ per gallon.

9. If the sales tax rate is 4%, how much tax would you owe on a coat that sells for $39.59?

10. How much would you pay for the coat in problem 9 including tax?

11. If a credit card company charges a monthly finance charge of 1.5%, what would be the monthly finance charge on $137?

12. Joan bought a toaster for $14.97. How much tax did she owe on the toaster if the tax rate in her state is 6%?

13. What was the total amount Joan, in problem 12, had to pay for the toaster including tax?

14. Find the price of 21 gallons of gasoline at 62.8¢ per gallon.

Finding Interest for One Year

Interest is money that money earns. On a loan, interest is the payment you must make for using the lender's money. On a savings account, interest is the money the bank pays you for using your money.

The formula for finding interest is $I = PRT$:

I is the **interest** in dollars

P is the **principal,** the money borrowed or saved

R is the percent **rate,** which can be written as either a fraction or a decimal

T is the **time** in years

The formula is read as: **Interest is equal to the principal times the rate times the time.**

EXAMPLE 1.   Find the interest on $500 at 8% annual interest for one year.

$$\frac{\overset{5}{\cancel{500}}}{1} \times \frac{8}{\underset{1}{\cancel{100}}} \times 1 = \$40$$

Notice that the percent was changed to a fraction with a denominator of 100.

EXAMPLE 2.   Find the interest on $600 at $4\frac{1}{2}$% annual interest for one year.

$$\frac{\overset{300}{\cancel{600}}}{100} \times \frac{9}{\underset{1}{\cancel{2}}} \times 1 = \frac{2700}{100} = \$27$$

Here we changed the $4\frac{1}{2}$% to an improper fraction $\left(\frac{9}{2}\right)$ and put a denominator of 100 under the principal.

EXAMPLE 3.   Find the interest on $400 at 8.5% for one year.

$$\begin{array}{r} \$400 \\ \times\ .085 \\ \hline 2\ 000 \\ 32\ 00 \\ \hline 34.00\cancel{0} \\ \times\ 1 \\ \hline \$34.00 \end{array}$$

Here we changed the 8.5% to a decimal (.085).

1. Manny borrowed $800 at 10% annual interest. How much interest did he owe in one year?

2. Deborah borrowed $450 at 15% annual interest. How much interest did she owe in one year?

3. Frank kept $200 in his savings account for one year. If his money earned $5\frac{1}{2}\%$ annual interest, how much interest did he get in a year?

4. Diane borrowed $1,000 in order to finish a year in school. If she had to pay 6.7% interest on the loan, how much interest did she owe after one year?

5. Ellen saved $600 for a year. How much interest did she earn if her bank pays an annual interest rate of $6\frac{1}{4}\%$?

6. How much interest did Daniel owe on a $1,500 loan in one year if his bank charged him 11.5% interest?

7. Jose had to pay 8.4% interest on a $950 loan. How much interest did he owe in one year?

8. Madge kept $1,200 in her savings account for a year. How much interest did she earn if her bank pays $4\frac{3}{4}\%$ annual interest?

Finding Interest for Less than One Year

Interest rates are usually given for one year. If you want to find interest for less than one year, write the number of months over 12 (the number of months in one year). Use this fraction for **T** in the formula I = PRT.

EXAMPLE: Find the interest on $500 at 8% annual interest for nine months.

Step 1.   9 months $= \dfrac{9 \text{ months}}{12 \text{ months per year}} = \dfrac{3}{4}$ year

Step 2. $\dfrac{\cancel{500}^{5}}{1} \times \dfrac{\cancel{8}^{2}}{\cancel{100}_{1}} \times \dfrac{3}{\cancel{4}_{1}} = \$30$

---

1. Find the interest on $900 at 6% annual interest for four months.

2. Mary's bank pays 4.5% annual interest on savings accounts. How much interest will Mary make on $1,000 in six months?

3. David's charge account costs him 18% in annual interest. How much interest would he owe on $700 in 10 months?

4. Jane kept $270 in her savings account for 8 months. How much interest will she get if her bank pays 4% annual interest?

5. Fred was charged 7.2% annual interest on a $1,200 loan. How much interest did he owe if he paid the loan back in two months?

Finding Interest for More than One Year

If you want to find interest for more than one year, write the total number of months over 12 (the number of months in one year). Use this improper fraction for **T** in the formula I = PRT.

EXAMPLE: Find the interest on $600 at 5% annual interest for one year and 6 months.

Step 1. 1 year and 6 months $= \dfrac{18 \text{ months}}{12 \text{ months}}$ years $= \dfrac{3}{2}$ years

Step 2. $\dfrac{\overset{3}{\cancel{600}}}{1} \times \dfrac{5}{\underset{1}{\cancel{100}}} \times \dfrac{3}{\underset{1}{\cancel{2}}} = \$45$

---

1. Find the interest on $800 at 9% annual interest for two years and three months.

2. Carlos borrowed $1,500 at 11% annual interest. He paid the loan back in one year and eight months. How much interest did he pay?

3. Sally kept $400 in her bank account for one year and nine months. How much interest did she make if her bank pays $4\frac{1}{2}$% annual interest?

4. Mr. Clay paid back his $2,000 car loan two years and six months after he borrowed the money. If he was charged 10.5% annual interest, how much interest did he pay?

5. Jack's bank pays $5\frac{1}{4}$% annual interest on Jack's account. How much interest will Jack make on a $600 deposit that he keeps in the bank for one year and four months?

Finding Compound Interest

In most savings accounts, money earns **compound interest.** This means that your balance (the principal) changes by the addition of interest on a regular time basis such as quarterly (every three months), monthly, or even daily.

EXAMPLE: Joe deposited $500 in his savings account. How much money will he have in the account at the end of a year if he gets 4% annual interest and the interest is **compounded** (added to his balance) quarterly.

Step 1. First quarter: $\dfrac{\overset{5}{\cancel{500}}}{1} \times \dfrac{\overset{1}{\cancel{4}}}{\underset{1}{\cancel{100}}} \times \dfrac{1}{\underset{1}{\cancel{4}}} = \$5.00$

Amount in account: $500 + $5 = $505.00

Step 2. Second quarter: Joe gets interest on the new balance, $505.00.

$\dfrac{\overset{101}{\cancel{505}}}{1} \times \dfrac{\overset{1}{\cancel{4}}}{\underset{20}{\cancel{100}}} \times \dfrac{1}{\underset{1}{\cancel{4}}} = \dfrac{101}{20} = \$5.05$

Amount in account: $505 + $5.05 = $510.05

Step 3. Third quarter: The interest is calculated on $510.05.

$\dfrac{\overset{102.01}{\cancel{510.05}}}{1} \times \dfrac{\overset{1}{\cancel{4}}}{\underset{20}{\cancel{100}}} \times \dfrac{1}{\underset{1}{\cancel{4}}} = \dfrac{102.01}{20} = \$5.10$ (rounded off to nearest cent)

Amount in account: $510.05 + $5.10 = $515.15

Step 4. Fourth quarter: The interest is calculated on $515.15.

$\dfrac{\overset{103.03}{\cancel{515.15}}}{1} \times \dfrac{\overset{1}{\cancel{4}}}{\underset{20}{\cancel{100}}} \times \dfrac{1}{\underset{1}{\cancel{4}}} = \dfrac{103.03}{20} = \$5.15$ (rounded off to nearest cent)

Amount in account: $515.15 + $5.15 = $520.30

Notice that every quarter the new principal was used to calculate the amount of interest.

In working these problems, round off your answers to the nearest cent.

1. Luis put $800 in his savings account, which pays 6% annual interest. How much will he have in his account in a year if the interest is compounded quarterly?

2. Denise kept $1,200 in an account that pays 8% annual interest. How much will she have in her account in a year if the interest is compounded quarterly?

Comparing Food Prices: Unit Pricing

Since the same food can come in different package sizes, it is not always easy to tell which package is the best buy. Suppose you find two different brands of peas, one in a 16-ounce can priced at 40¢ and one in a 10-ounce can priced at 35¢. You could figure out which one is the best buy by finding the price per ounce of peas in each can. Finding this price is called **unit pricing**.

EXAMPLE: Which of these two brands of peas is cheaper?

16-ounce can of Brand X peas — 40¢
10-ounce can of Brand Y peas — 35¢

The cost per ounce for Brand X is:

$$2\frac{8}{16} = 2\frac{1}{2}¢ \text{ per ounce}$$
$$16\overline{)40¢}$$
$$\underline{32}$$
$$8$$

The cost per ounce for Brand Y is:

$$3\frac{5}{10} = 3\frac{1}{2}¢ \text{ per ounce}$$
$$10\overline{)35¢}$$
$$\underline{30}$$
$$5$$

Brand X is cheaper.

Use the advertisements below to answer the questions on page 125.

FRED'S FOODS	GERT'S GROCERIES	SAM'S STORE
Margarine 2 8-oz. cups .69	Margarine ..... 3 8-oz. cups 99¢	**Margarine** 1-lb. pkg. **59¢**
48 size, Indian River seedless grapefruit **7** $**1** for	INDIAN RIVER SEEDLESS **GRAPEFRUIT** (48 size) **6** FOR **89¢**	FLORIDA SEEDLESS Grapefruit 4 for 49¢
orange juice ½-gallon cont. **69¢**	**Orange Juice** 64-OZ. BTL. **59¢**	**Orange Juice 3** 1 qt. conts. **$1⁰⁰**

What is the price of 8 ounces of margarine

1. at Fred's Foods?

2. at Gert's Groceries?

3. at Sam's Store? (1 pound = 16 ounces.)

4. Which store has the cheapest margarine?

What is the price of one seedless grapefruit

5. at Fred's Foods?

6. at Gert's Groceries?

7. at Sam's Store?

8. Which store has the cheapest grapefruit?

What is the price of one quart (32 ounces) of orange juice

9. at Fred's Foods?

10. at Gert's Groceries?

11. at Sam's Store?

12. Which store has the cheapest orange juice?

Finding the Percent Saved at a Sale

To find what percent you save by buying an item on sale, put **the amount saved over the original price**, reduce, and change the fraction into a percent.

EXAMPLE:

All the prints that fit...on sale! Long-sleeved blouses in assorted prints. 100% polyester, sizes 8-18.
reg. $15.......................**sale $10**

What percent of the original price can you save by buying the blouse on sale?

Step 1. The regular price is $15. The sale price is $10.
The amount saved is $15 − $10 = $5.

Step 2. $\dfrac{\text{savings}}{\text{original}} = \dfrac{\$5}{\$15} = \dfrac{1}{3} = 33\frac{1}{3}\%$

Find the percent of savings for each item shown in this advertisement.

STARR'S DISCOUNT STORE

Let Us Make You A Starr!

Gigantic Savings——Storewide Clearance Sale

Wizard® canister vacuum. 2.3 hp (peak-rated by manufacturer); automatic cord rewind; tools, more! Add $2 deliv.
orig. $100 .... **sale $62**

Fashion's most popular pre-washed denim. 45" wide.
reg. $4 yd. .... **sale $2 yd.**

Men's jackets with the rich look of leather. Zip or button-front. S-M-L-XL. reg. $35 .... **sale $20**

Misses' acrylic rib crewneck pull-over for skirts, slacks, dresses! Stripe combinations of blue, green brown. S,M,L.
reg. $11 .... **sale $8**

Men's famous maker brushed cotton jean leisure sets. Buy separates or set. Faded blue or bone. Snap-front jacket, sizes S-M-L-XL.
reg. $27 .... **sale $19**

Men's long sleeve print sport shirts. Splashy patterns, geometrics; assorted colors. Cotton, polyester/cotton or nylon. S-M-L-XL.
orig. $15 .... **sale $8**

Flared slacks, sizes 32-42.
orig. $18 .... **sale $9**

Flared four-pocket jeans, sizes 32-40R, L. reg. $18 .... **sale $14**

Boys' famous-maker corduroy jeans, slacks. Assorted styles, many colors in durable-press polyester/cotton, 4-7R, S.
reg. 7.50 .... **sale $6**

Geometric-design comforter. Lightweight; cotton/polyester with polyester fill. Brown/blue reverses to solid blue.
Twin, orig. $25 .... **sale $19**

Full, orig. $35 .... **sale $29**

1. Wizard vacuum—percent saved _____

2. Pre-washed denim—percent saved _____

3. Men's jackets—percent saved _____

4. Misses' crewneck pullover—percent saved _____

5. Men's jean leisure suit—percent saved _____

6. Men's long-sleeved sport shirt—percent saved _____

7. Flared slacks—percent saved _____

8. Flared four-pocket jeans—percent saved _____

9. Boys' corduroy jeans—percent saved _____

10. Twin-size comforter—percent saved _____

11. Full-size comforter—percent saved _____

Buying Furniture on Sale

When Deborah Robinson went into The Friendly Furniture Store on February 3rd, she saw these signs:

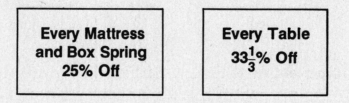

Every Mattress and Box Spring 25% Off

Every Table 33⅓% Off

Deborah decided to buy a mattress and box spring which had originally sold for $78 each and a kitchen table which originally cost $49.95.

1. How much money could she save on the mattress?

2. What was the sale price of the mattress?

3. How much money could she save on the kitchen table?

4. What was the sale price of the kitchen table?

5. What was the total price for the mattress, the box spring, and the kitchen table?

6. In Deborah's state there is an 8% sales tax. How much did she owe for her furniture including tax?

7. On February 4th, Deborah deposited her paycheck of $135.70 and $30 in cash in her checking account. Fill out the deposit slip on the next page as she would.

For credit to the Account of

DEBORAH ROBINSON

Bank Use Only

Chk		
Less Dep		
Ret'd		

Units

Date:

CHEMICAL BANK

Checking Account Deposit

P&R 63 (7-71)

	Dollars	Cents
Cash Include Coupons		
Checks List Each Check 1		
2		
3		
4		
5		
6		
7		
8		
Total		

RC

8. On February 6 of this year, Deborah went back to the store and paid for the mattress, box spring, and kitchen table with a check. Fill out the check below as she would and enter the check and deposit on the form that follows.

No. **004**

_____ 19___ **1-12/210**

PAY TO THE ORDER OF _____ $_____

_____ DOLLARS

CHEMICAL BANK

FOR _____

CHECK NO.	DATE	CHECK ISSUED TO	AMOUNT OF CHECK	DATE OF DEP.	AMOUNT OF DEPOSIT	BALANCE
						285 00
001	1/21	Telephone Company	35 70			249 30
002	1/29	Utility Company	15 36			233 94
003	2/1	Rental Agent	145 00			88 94
004						

Using a Tax Rate Schedule

An important part of filling out income tax forms is using a tax rate schedule. Tax rate schedules let you know how much money you have to pay in taxes based on how much money you made during the year. The tax rate schedule on this page is for heads of household. The **head of household** is the person who brings in the money or most of the money on which a family lives.

This schedule is used to figure out the amount of money owed to the federal government. The amount is based on the taxable income entered on line 5 of a form called the Estimated Tax Worksheet. Both Tax Rate Schedules and Estimated Tax Worksheets can be obtained from the Internal Revenue Service.

Schedule Z—Unmarried or legally seprated Taxpayers Who Qualify as Heads of Household

If the amount on line 5, Estimated Tax Worksheet, is:		Enter on line 6, Estimated Tax Worksheet:	
Not over $1,000		14% of the amount on line 5	
Over—	But not over—		of excess over—
$1,000	$2,000	$140+16%	$1,000
$2,000	$4,000	$300+18%	$2,000
$4,000	$6,000	$660+19%	$4,000
$6,000	$8,000	$1,040+22%	$6,000
$8,000	$10,000	$1,480+23%	$8,000
$10,000	$12,000	$1,940+25%	$10,000
$12,000	$14,000	$2,440+27%	$12,000
$14,000	$16,000	$2,980+28%	$14,000
$16,000	$18,000	$3,540+31%	$16,000
$18,000	$20,000	$4,160+32%	$18,000
$20,000	$22,000	$4,800+35%	$20,000
$22,000	$24,000	$5,500+36%	$22,000

EXAMPLE: What amount goes on line 6 of the Estimated Tax Worksheet if the amount on line 5 is $4,570?

Step 1. Read the Tax Rate Schedule from left to right. Since $4,570 is between $4,000 and $6,000, the amount for line 6 is $660 plus 19% of the excess (additional dollars) over $4,000.

Step 2. The excess over $4,000 is:

$$\begin{array}{r} \$4,570 \\ -\ 4,000 \\ \hline 570 \end{array}$$

Step 3. 19% of $570 is:

$$\begin{array}{r} \$570 \\ \times\ .19 \\ \hline 51\ 30 \\ 57\ 0 \\ \hline 108.30 \end{array}$$

Step 4.

$$\begin{array}{r} \$660.00 \\ +\ 108.30 \\ \hline 768.30\ \text{tax owed} \end{array}$$

Use the schedule on page 130 to find the following.

1. If line 5 is $1,380, the amount on line 6 will be _____.

2. If line 5 is $2,695, the amount on line 6 will be _____.

3. If line 5 is $5,780, the amount on line 6 will be _____.

4. If line 5 is $6,040, the amount on line 6 will be _____.

5. If line 5 is $7,508, the amount on line 6 will be _____.

6. If line 5 is $10,430, the amount on line 6 will be _____.

7. If line 5 is $12,260, the amount on line 6 will be _____.

8. If line 5 is $15,870, the amount on line 6 will be _____.

9. If line 5 is $17,050, the amount on line 6 will be _____.

10. If line 5 is $20,692, the amount on line 6 will be _____.

Filling Out a Wage and Tax Statement

Peter Munro made a gross salary (before deductions) of $9,650 in 1981 at Coljer Products in New York. His employers withheld (deducted) from his gross pay certain amounts for federal income tax, state income tax, city income tax, and FICA (social security).

1. 17.7% of Mr. Munro's gross annual salary of $9,650 was withheld for federal income tax. How much money was withheld from his pay for federal tax?

2. 5.8% of Mr. Munro's gross salary was withheld for FICA. How much money was withheld for FICA?

3. 4.1% of Mr. Munro's gross salary was withheld for state income tax. How much money was withheld for state tax?

4. 1.4% of Mr. Munro's gross salary was withheld for city income tax. How much was withheld for city tax?

5. What was Mr. Munro's net (take-home) pay for 1981?

6. Fill in the Wage and Tax Statement below. Put Mr. Munro's gross pay in boxes 2, 4, 7, and 10. Put the federal income tax withheld in box 1. Put the FICA withheld in box 3. Put the state tax withheld in box 6. Put the city tax withheld in box 9.

COLJER PRODUCTS
333 EAST 49TH STREET
N.Y., N.Y. 10017   13-5009340

Wage and Tax Statement 1981

Type or print EMPLOYER'S Federal identifying number, name, address, and ZIP code above.

| Employer's State identifying number 21-49521 | Copy A For Internal Revenue Service Center |

FEDERAL INCOME TAX INFORMATION		SOCIAL SECURITY INFORMATION		STATE OR LOCAL INCOME TAX INFORMATION			
1 Federal income tax withheld	**2** Wages, tips, and other compensation	**3** FICA employee tax withheld	**4** Total FICA wages	**6** Tax withheld	**7** Wages paid	**8** State or locality NEW YORK	

EMPLOYEE'S social security number ▶ 123-45-6789

| **5** Uncollected employee FICA tax on tips NONE | **9** Tax withheld | **10** Wages paid | **11** State or locality NEW YORK |

PETER MUNRO
5 EAST THIRD STREET
N.Y., N.Y. 10003

OTHER INFORMATION				STATUS
Was employee covered by a qualified pension plan etc.? Yes ☐ No ☐	Contribution to individual employee retirement account	Cost of group term life insurance included in box 2	Excludable sick pay included in box 2	1. Single 2. Married

If this is a corrected form, put an "X" to the right of the number in the upper left corner.

Type or print EMPLOYEE'S name, address, and ZIP code above.

For instructions see Form W-3 and back of Copy D.

Form **W-2**

Department of the Treasury—Internal Revenue Service

7. Mr. Munro owed the federal government $1,815 for 1981. How much greater is $1,815 than the amount that was withheld from his pay for federal income tax?

8. Fill in the check below to the Internal Revenue Service as Mr. Munro would on April 15, 1982, for the balance that he owed.

No. 021

_____ 19____   1-12/210

PAY TO THE ORDER OF _____ $ _____

_____ DOLLARS

CHEMICAL BANK

FOR _____   _____

Working with a Budget

The budget below shows the way that the four members of the Johnson family spend their money.

Johnson Family Budget			
Rent	20%	Clothes	10%
Food	31%	Medical	12%
Utilities	5%	Savings	10%
(Electricity, Gas, Telephone)		Odds and Ends	6%
Entertainment	6%		

1. The income for the Johnson family comes from Mr. Johnson's yearly salary of $10,500. 21% of his salary is withheld for taxes and social security. How much money is withheld for taxes and social security in a year?

2. What is Mr. Johnson's yearly take-home pay?

For the following questions use the budget above and Mr. Johnson's yearly **take-home** pay.

3. How much do the Johnsons pay every year for rent?

4. How much do the Johnsons pay each month for rent?

5. How much do the Johnsons spend in a year
    for food?

6. To the nearest cent, how much do the
    Johnsons spend each week for food?
    (One year = 52 weeks.)

7. How much do the Johnsons spend in a year
    for utilities?

8. How much do the Johnsons spend in a year
    for entertainment?

9. How much do the Johnsons spend in a year
    for clothes?

10. There are four people in the Johnson family.
    To the nearest cent, what is the average spent
    on each person in the family for clothes in
    one year?

11. How much do the Johnsons spend in a year
    for medical expenses?

12. How much do the Johnsons save in one year?

13. How much money goes for odds and ends in
    one year?

Buying on the Installment Plan

Buying on the installment plan is an easy way to buy things without paying the entire amount in cash. However, by paying in regular installments, a customer can end up paying much more than the cash price of an item. The problems on the next three pages show how different cash prices and installment prices can be.

1. A table and chair set marked $159 can be bought for 10% down and $15 a week for twelve weeks.

   (a) What is the cash price of the table and chair set including an 8% sales tax?

   (b) How much does the table and chair set cost on the installment plan?

   (c) How much more does it cost to buy the table and chair set on the installment plan than it does to pay cash?

2. A set of living room furniture marked $395 can be bought for 15% down and $35 a month for a year.

    (a) What is the cash price of the furniture including a 6% sales tax?

    (b) How much does the furniture cost on the installment plan?

    (c) How much more does it cost to buy the furniture on the installment plan than it does to pay cash?

3. A color television marked $459 can be bought for 8% down and $10.50 a week for a year.

    (a) What is the cash price of the television including a 5% sales tax?

    (b) How much does the television cost on the installment plan?

    (c) How much more does it cost to buy the television on the installment plan than to pay cash?

4. A set of bedroom furniture marked $289.95 can be bought for 12% down and $14.80 a week for 22 weeks.

    (a) What is the cash price of the furniture including a 6% sales tax?

    (b) How much does the furniture cost on the installment plan?

    (c) How much more does it cost to buy the furniture on the installment plan than to pay cash?

5. An air conditioner marked $114.50 can be bought for 16% down and $12.50 a month for a year.

    (a) What is the cash price of the air conditioner including a 5% sales tax?

    (b) How much does the air conditioner cost on the installment plan?

    (c) How much more does it cost to buy the air conditioner on the installment plan than to pay cash?

pages 2-4

1. 5/8
2. 2/3
3. 2 2/5
4. 38/7
5. 8 1/5
6. 11 11/12
7. 26 13/16 pounds
8. 6 1/4 hours
9. 6 7/20
10. 4 1/2
11. 3 13/21
12. 124 1/2 pounds
13. 27 1/2 inches
14. 21/40
15. 1/4
16. 15
17. 28 crates
18. 37 1/2 hours
19. 1 1/9
20. 16
21. 1/18
22. 2 3/16
23. 28 bags
24. 10 pieces

page 5

1. 1/6
2. 5/8
3. 3/4
4. 3/8
5. 2/5
6. 4/9
7. 5/6
8. 2/3
9. 5/9

page 6

1. 5/12
2. 47/100
3. 9/16
4. 23/36
5. 7/12
6. 8/25
7. 43/60
8. 1351/2000
9. 2/5
10. 63/100
11. 3/4
12. 113/150
13. 77/280

page 7

1. I P M I
2. P I I M
3. M P P I
4. P I M P

page 9

1. 1/2   1/4   1/3   1/5   1/8
2. 5/6   8/9   8/9   7/8   2/3
3. 2/5   1/3   7/20   9/14   8/17
4. 3/7   3/4   1/2   3/4   3/4
5. 15/16 7/8   4/5   2/3   9/16
6. 1/2   1/3   1/50   7/9   5/7

page 10

1. 24/30 18/20 3/18   20/32
2. 20/35 18/36 14/21 54/66
3. 25/45 33/44 35/60 15/45

page 11

1. 1 3/4   5 1/2   2 4/5   4 2/7   4
2. 3 1/3   3 1/4   3   3 3/5   2
3. 1 1/12   5   7 1/2   2 2/3   4 2/3

page 12

1. 11/4   11/7   16/3   44/7   23/5
2. 19/2   61/8   29/10   35/4   32/9
3. 31/3   57/5   53/12   55/8   49/4

pages 13-15

1. 5/9   4/7   7/8   7/12   10/13
2. 6/11   8/9   13/15   15/17   16/19
3. 7 3/5   14 9/10   9 9/11   14 11/13
4. 12 8/9   17 9/11   20 6/7   15 9/10
5. 3/4   2/3   2/3   2/3   1/2
6. 4/5   4/5   7/10   3/4   2/3
7. 11 1/2   16 2/3   18 6/7   42 2/3
8. 24 2/3   17 4/5   17 3/4   31 3/4
9. 1 2/5   1 3/8   1 3/10   1 4/9   1
10. 1 1/3   1 1/7   1 1/2   1 1/5   1 1/2
11. 2 2/9   2   1 1/2   2 1/4   2 1/4
12. 17 7/8   18 2/3   18 1/10   18

pages 16-19

1. 1 1/4   1 1/2   1 5/8   1 1/6   1 2/9
2. 1 5/8   1 1/3   1 4/5   1 1/3   1 1/5
3. 1 2/5   2   11/12   1 2/3   1 2/5
4. 22 1/3   14 1/6   15 1/8   22 1/6
5. 21   18 1/16   17 1/5   17
6. 1 4/15   1 1/12   9/10   16/21   1 7/30
7. 1 9/28   1 5/42   1 7/40   1 1/33   1 7/45
8. 1 7/12   1 5/18   1 9/20   35/36   1 5/24
9. 2 1/24   1 11/20   2 17/36   1 7/16   1 5/9
10. 1 1/2   1 23/24   1 3/4   1 1/40   1 1/36
11. 1 17/18   2 3/28   13/18   1 52/63   1 43/48
12. 11 7/20   17 7/24   12 2/9   13 19/24
13. 18 13/21   10 11/12   16 3/20   16 5/18
14. 13 5/24   11 1/18   15 19/40   12 17/30
15. 17 29/30   19 7/24   17 29/36   22 1/12
16. 13 1/24   21 19/36   15 9/16   23 1/4
17. 15 11/12   24 11/70   24 11/18   17 45/56

page 20

1. 68 1/4 inches   2. 16 1/10 miles
3. 22 7/20 pounds   4. 8 19/24 pounds
5. 14 1/4 hours   6. 128 7/10 pounds
7. 2 23/30 hours

page 21

1. 1/3   1/10   1/2   3/13   6/11
2. 1/3   3/8   1/2   3/19   1/5
3. 3 4/7   6   1 1/3   4 2/5
4. 8 1/4   9 1/2   9 5/13   16 2/3

pages 22-23

1. 1/4   3/8   1/2   9/16   1/5
2. 5/12   7/15   13/28   7/30   7/18
3. 6 13/24   6 3/14   7 26/45   3 1/20
4. 6 11/56   15 23/36   7 13/30   12 5/33
5. 5 1/4   3 3/8   2 2/21   2 5/16   9 1/5
6. 5 14/45   1 1/28   7 1/30   7 7/24   4 1/12
7. 7 17/56   7 11/63   1 23/72   4 17/42   2 11/24
8. 6 5/24   3 1/30   2 7/20   8 1/12   8 1/30
9. 4 1/14   2 5/24   7 2/15   9 3/10   6 1/6
10. 4 1/10   11 13/84   3 7/30   5 1/9   3 9/40

pages 25-26

1. 7 1/6   3 4/7   11 1/2   8 3/5   9 3/11
2. 3 4/7   3 1/3   1 5/12   3 4/9   6 11/16
3. 3 2/3   6 1/2   7 2/3   6 2/5
4. 4 14/15   15 5/7   6 2/3   7 3/4
5. 10 1/2   27 8/11   7/10   2 2/3
6. 5 13/20   4 5/8   4 7/9   14 5/12
7. 18 13/14   8 7/12   5 4/5   31 38/45
8. 8 19/42   7 23/40   13 20/33   1 7/12
9. 23/30   3 25/36   13 43/56   6 17/24
10. 10 17/30   6 7/8   13 5/6   17/22
11. 1 37/56   16 13/20   5 7/18   14 25/48

page 27

1. 20 7/8 inches
2. 154 1/4 pounds
3. 489 2/5 miles
4. 3 5/6 pounds
5. 8 5/8 pounds
6. 11 2/3 rpm's
7. 55 3/8 pounds
8. 1/8 pound

page 28

1. 8/15   10/63   7/80   15/88
2. 1/15   16/63   25/48   9/40
3. 14/45   21/64   5/36   16/81
4. 9/40   5/42   10/81
5. 16/75   14/135   8/63

pages 29-30

1. 3/10   4/21   7/16   5/14
2. 2/15   16/39   45/77   4/65
3. 1/6   3/8   1/4   3/4
4. 9/20   4/15   3/8   5/18
5. 3/10   8/11   2/5   3/20
6. 9/52   2/7   5/7   1/3
7. 15/32   1/9   8/15
8. 3/8   7/60   3/32
9. 4/45   9/100   2/9
10. 25/2016   7/425   19/120
11. 55/567   14/99   1/16

page 31

1. 1 5/7   2 1/4   6 2/3   2 2/5
2. 10   10   1 1/3   24
3. 21   5 1/2   14   11 1/4
4. 8 1/6   3 1/3   21   1 4/5

page 32

1. 3/8   10/21   1 31/32   1 13/20
2. 1 2/3   3/4   4 3/8   5 1/7
3. 2 4/5   25   16 1/4   38 1/2
4. 4   22 1/2   84

page 33

1. 32 hours   2. 184 miles
3. 200 pounds   4. 171 inches
5. $11   6. $13
7. $4.50   8. 9 1/2 yards

page 35

1. 1 1/14   2/3   11/24   7/12
2. 1 1/10   4   28/39   1 1/5
3. 1 23/40   1 4/45   1 2/7   1 1/5
4. 3/28   4   3 1/30   2 2/15
5. 35/66   1 29/49   8 1/2   3/16
6. 7/20   3/5   1 1/3   1 1/63

page 37

1. 30   27   6 2/3   9
2. 21 1/3   20   34   36
3. 50   84   37 1/2   66
4. 90   36   26 2/3   31 1/2
5. 168   47 1/2   16 1/2   13 1 3
6. 32   5 1/3   8 1/3   58 1/3

page 39

1. 1/5 1/9 2/19 1/8
2. 1/36 3/28 1/12 4/13
3. 1/22 1/25 1/8 2/39
4. 5/126 1/40 2/75 3/125
5. 1/15 1/39 1/51 3/88
6. 1/77 1/48 1/54 1/120

pages 40-41

1. 2 2 1/2 4 2/5 19 1/2
2. 2/5 5/6 4/21 4/27
3. 1/2 4/5 7/30 3/20
4. 7 1/2 8 3/4 4 4/5 14 2/3
5. 3 1/3 3 6/7 1 7/15 4/5
6. 1 1/8 2 1 19/30 2 4/5
7. 3 1/2 1 19/21 4
8. 2 2/3 3 2 1/4 1 1/39
9. 2 1/45 1 13/42 5 7/18 1 43/61
10. 2 11/32 1 23/41 1 29/41 2 13/36

page 42

1. 6 pieces   2. 1 1/2 pounds   3. 6 suits
4. 6 loaves   5. 32 cans   6. 9 strips
7. 4 dresses (with 1 1/2 yards of material left over)

pages 43-45

1. 7/12
2. 9/16
3. 4 1/6
4. 93/8
5. 19 7/8
6. 20 23/30
7. 2 5/12 hours
8. 13 5/16 pounds
9. 5 19/40
10. 5 6/11
11. 8 7/18
12. 3/4 pound
13. 1 5/6 miles
14. 20/63
15. 2/5
16. 3 1/3
17. 65 pounds
18. 96 ¢
19. 32/35
20. 31 1/2
21. 1/26
22. 15/28
23. 6 bookcases
24. $4

pages 46-48

1. .076
2. 3 1/25
3. .41 2/3
4. .08
5. .07, .08, .087, .7
6. 4.439
7. 23.953
8. 12.1 pounds
9. 84.1°
10. 9.143
11. 11.63
12. 1.15 yards
13. 286.8 miles
14. 10.971
15. .0228
16. 49.2
17. 144 miles
18. $26.10
19. .86
20. 7.3
21. 370
22. 200
23. .496
24. $1.80
25. 5 yards

page 50

1. eight tenths   three tenths
2. six hundredths   thirty-seven hundredths
3. nine hundredths   five thousandths
4. twenty-three thousandths
   twenty-eight ten-thousandths
5. three hundred seventy-eight thousandths
   six ten-thousandths
6. forty-one ten-thousandths
   three thousand five hundred seventy-seven
      ten-thousandths
7. twenty-eight and sixteen hundredths
   fifty-six and twenty-five thousandths
8. four and eight thousandths
   ten and three hundred seventy-five **thousandths**
9. sixty-three and seventy-eight ten-thousandths
   nineteen and twenty-eight hundred-thousandths
10. eight and three hundred twenty-six millionths
   two and three millionths

page 51

1.	.7	12.103
2.	4.9	250.6
3.	.006	.0321
4.	.022	399.5
5.	.0015	60.312
6.	.0085	20.00009
7.	4.07	.00318
8.	.000019	10,000.0010

page 52

1. 2/25   3/8   3/625
2. 3 3/5   9 43/50   10 1/500
3. 17/200   5 2/25   1/400
4. 81/25,000   19 393/5,000   123 231/500
5. 16 1/25,000   7 11/50   3 1/125,000
6. 2,036 4/5   48 1/50   3 3/40

page 53

1. .25   .4   .625 or .62 1/2   .33 1/3
2. .22 2/9   .24   .16 2/3   .375 or .37 1/2
3. .83 1/3   .3   .57 1/7   .41 2/3

page 54

1.	.04		2.	.99
3.	.33		4.	.11
5.	.006		6.	.4

7. .03, .033, .303, .33   8. .08, .082, .28, .8
9. .061, .106, .16, .6   10. .007, .017, .02, .2
11. .045, .4, .405, .45   12. .04, .304, .32, .4
13. .0072, .02, .027, .07   14. .026, .06, .0602, .2

page 55

1. 1.207   2. 25.18   3. 134.64
4. 39.617   5. 24.313   6. 65.303
7. 75.449   8. 18.8097   9. 132.543
10. 93.277

page 56

1. 10.66 inches   2. 1200 miles   3. 76.9 inches
4. 9.7 pounds   5. 121.5 miles   6. 103.1°
7. 7.02 kilograms   8. 210.4 million people

page 57

1. 0.44   2. 0.6367   3. 0.52
4. 0.122   5. 54.36   6. 88.534
7. 11.064   8. 0.953   9. 12.068
10. 7.769   11. 0.068   12. 2.507
13. 88.515   14. 3.0014   15. 0.0708

page 58

1. 7.8 years   2. 106.4 people   3. 0.33 meters
4. 3.1 miles   5. 402.7 miles   6. 123.15 meters
7. 26.5 million phones   8. 6.4¢

pages 59-61

1. 15.2   8.28   40.2   42.4   5.88
2. 1.23   39.0   5.31   34.4   1.14
3. 277.6   20.23   3.306   542.7   28.72
4. 63.42   1.008   626.4   39.12   1.401
5. 142.50   27.720   1341.0   560.70   550.20
6. .054   .035   .0012   .0032   .0016
7. 1.12   .0584   6.44   .0348   .198
8. 29.26   .3120   .2580   4.428   1.388
9. 11.70   2.496   8.188   2.414   .2184
10. .486   2.184   26.46   .1411   6.072
11. 12.04   4.592   4.104   .5589   1.692
12. 8.235   48.840   296.64   .46774   66.138
13. 253.176   273.604   57.932   3.66912
14. 7252.3   36.2142   72.8900   3.44787
15. 2.7712   .14904   44.7468   836.368
16. .047413   112.2740   8.29376   17,210.5704

page 62

1.   8   .9   36.4   7.21
2.   3   27.5   890   86.3
3.   900   2360   475   1600
4.   3.4   124   3850   60

page 63

1.   30.48 cm.   2. $28.50   3.   40 km.
4.   106.15 miles   5.   72 kg.   6.   1000 pounds
7. $11.27   8. $14.45   9.   68.55 meters
10.   3.78¢

page 64

1.   2.3   7.8   .096   8.25
2.   .024   49.6   3.89   .631
3.   4.8   .36   .08   5.7
4.   .183   36.4   6.03   .517

pages 65-66

1.   9.6   38   .7   6.9
2.   .28   14   .83   .09
3.   14.7   1.93   605   12.8
4.   .06   2.07   40.4   .182
5.   760   480   2700   9640
6.   13,300   4   720   85
7.   325   71,400   935   2080
8.   490   3070   560   9310

page 67

1.   70   400   3700   6400
2.   9000   8000   3900   6280
3.   460   7200   500   420
4.   62,000   78,000   5900   83,000

page 68

1.   .09   3.6   2.73   .004
2.   .142   .013   7.28   .006
3.   .0375   .0018   .002   .428
4.   1.345   .0032   6.954   .0158

page 69

1.   127.2 pounds   2.   249.5 miles   3. $4.80
4.   38.5 hours   5.   384 mph   6. $1.60
7.   23 miles   8.   18 inches

pages 70-72

1.	.0014	**2.**	2/125
3.	.4375	**4.**	.02
5.	.013, .03, .031, .1	**6.**	21.356
7.	9.6973	**8.**	23.1 thousand
9.	11.35 pounds	**10.**	12.203
11.	2.346	**12.**	1.984 inches
13.	2.51 million	**14.**	2.844
15.	710.736	**16.**	2080
17.	16.51 centimeters	**18.**	$6.75
19.	4.09	**20.**	7.2
21.	6	**22.**	250
23.	.00513	**24.**	34.5 mph
25.	$3.80		

pages 73-75

1.	1.7%	**2.**	.04	**3.**	62 1/2%
4.	3/25	**5.**	1/12	**6.**	12
7.	11.5	**8.**	36	**9.**	$0.36
10.	$264	**11.**	33 1/3%	**12.**	8%
13.	11%	**14.**	66 2/3%	**15.**	80
16.	56	**17.**	$640	**18.**	$6.50

page 76

1. 32%   9%   60%   13.6%
2. .5%   37 1/2%   8 1/3%   4.5%
3. .16%   .03%   2.5%   33 1/3%
4. 12.5%   3.75%   .9%   80%

page 77

1. .2   .35   .08   .6
2. .035   .004   .0003   .216
3. .62 1/2   .06 2/3   .028   .19

50% = .50	5% = .05	37.5% = .375
25% = .25	1% = .01	62.5% = .625
75% = .75	100% = 1.00	87.5% = .875
20% = .2	10% = .1	12.5% = .125

page 78

1. 40%   25%   33 1/3%   37 1/2%
2. 24%   66 2/3%   83 1/3%   12 1/2%
3. 90%   87 1/2%   55%   41 2/3%
4. 16 2/3%   80%   70%   8 1/3%
5. 62 1/2%   44 4/9%   42 6/7%   45%
6. 16%   30%   60%   18%

page 79

1. 7/20   1/5   1/8   3/50
2. 1/6   1/100   9/10   3/8
3. 3/25   99/100   2/3   9/200
4. 4/5   1/3   1/25   1/12

page 80

1/2 = 50%   1/8 = 12 1/2 %
1/4 = 25%   3/8 = 37 1/2 %
3/4 = 75%   5/8 = 62 1/2 %
                        7/8 = 87 1/2 %
1/5 = 20%   1/10 = 10%
2/5 = 40%   3/10 = 30%
3/5 = 60%   7/10 = 70%
4/5 = 80%   9/10 = 90%
1/3 = 33 1/3 %   1/6 = 16 2/3 %
2/3 = 66 2/3 %   5/6 = 83 1/3 %

page 81

1. 6   67.55   78
2. 7.2   7.5   24
3. 510   400   209
4. 168   403   799
5. 10.14   .448   13.554

page 82

1. 25   12   14
2. 270   525   3
3. 21   60   175
4. 3   3 3/4   110

pages 83-84

1. 24 days   2. $1750   3. $9.60
4. $0.15   5. 6 answers   6. 351 members
7. $28.80   8. 6 times   9. $9.44
10. $20   11. 9.75 pounds   12. $1,275
13. $4.20   14. $56   15. $300
16. 10 days

pages 85-86

1. 50%   25%   33 1/3 %
2. 10%   16 2/3 %   12 1/2 %
3. 80%   40%   66 2/3 %
4. 83 1/3 %   75%   87 1/2 %

5. 87 1/2 %   90%   80%
6. 7%   6%   9%
7. 35%   38%   32%
8. 58 1/3 %   100%   44 4/9 %
9. 83 1/3 %   42 6/7 %   90%
10. 2%   6%   8 1/3 %
11. 20%   80%   22 1/2 %
12. 16 2/3 %   26%   .396%

pages 87-88

1. 75%   2. 8%   3. 62 1/2%
4. 90%   5. 8%   6. 5%
7. 33 1/3 %   8. 25%   9. 55%
10. 62 1/2 %   11. 8 1/3 %   12. 25%
13. 12 1/2 %   14. 5%   15. 16 2/3%
16. 8 1/3 %

pages 89-90

1. 32   90
2. 64   125
3. 150   80
4. 63   24
5. 75   1280
6. 1800   1400
7. 64   180
8. 480   240
9. 192   1500
10. 305   1050
11. 438   875

pages 91-92

1. $160   2. 30 problems
3. $2500   4. $168
5. $60   6. 120 members
7. $1700   8. $34,000
9. 225 pounds   10. 18,000 people
11. 3000 miles   12. $2000
13. $0.80   14. $194

pages 93-95

1. 12.3%   2. .056   3. 58 1/3 %
4. 21/25   5. 5/16   6. 24.3
7. 51.98   8. 6   9. $24.75
10. 9,000 people   11. 62 1/2 %   12. 8 1/2 %
13. 75%   14. 62 1/2 %   15. 200
16. 384   17. 420 miles   18. 32 letters

pages 96-100

1. 4/9
2. 20 1/24
3. 59 1/6 inches
4. 4 17/36
5. 6 15/16 inches
6. 1/15
7. 40 1/2
8. $15.75
9. 1/25
10. 4 1/5
11. 48 bags
12. 50.208
13. .02, .021, .12, .2
14. 309.799
15. 128.1 pounds
16. 39.613
17. 136.2 miles
18. 3.8912
19. 5.392
20. $3.60
21. .26
22. 19.3
23. 15 miles
24. .079
25. 56 1/4 %
26. 1/6
27. 54.72
28. 12
29. $110.72
30. 83 1/3 %
31. 62 1/2 %
32. 160
33. $250,000

page 102

1. C — 1 5/8 inches   H — 3 7/16 inches
   D — 2 inches   I — 3 15/16 inches
   E — 2 1/4 inches   J — 4 1/2 inches
   F — 2 7/8 inches   K — 4 13/16 inches
   G — 3 1/8 inches   L — 5 9/16 inches
2. 3/8 inch   3. 9/16 inch   4. 1 1/16 inches

page 103

1. C — 2.9 centimeters   H — 7.5 centimeters
   D — 3.4 centimeters   I — 10.4 centimeters
   E — 4 centimeters   J — 11.7 centimeters
   F — 5.5 centimeters   K — 13.2 centimeters
   G — 7.1 centimeters   L — 15 centimeters
2. 1.1 centimeters   3. 1.6 centimeters
4. 1.5 centimeters

pages 104-105

1. 21 1/3 inches
2. 18 1/4 feet
3. 39 inches
4. 24 meters
5. $108
6. 185 1/2 inches
7. 29.8 centimeters
8. 25 yards
9. 10 inches
10. 7 meters

pages 106-107

1. 20 sq. ft.
2. 17.36 sq. cm.
3. 30 sq. in.
4. 39.6 sq. m.
5. $253.44
6. 7 1/2 sq. yd.
7. 12.96 sq. m.
8. 5 4/9 sq. ft.
9. 12 1/4 sq. in.

pages 108-109

1. 10 cu. ft.
2. 531.25 cu. ft.
3. 918 cu. ft.
4. 1500 cu. in.
5. 7 1/2 cu. in.
6. 200 boxes
7. 1279.08 cu. m.
8. 84 cu. ft.
9. 26 cu. ft.

pages 110-111

1. 176 yards
2. 110 feet
3. $93.50
4. 31 3/7 inches
5. 11 feet
6. $3.19
7. 88 inches
8. 44 inches
9. 94 2/7 feet

pages 112-113

1. 154 sq. ft.
2. 314 2/7 sq. ft.
3. 616 tiles
4. $12.32
5. 13.86 sq. m.
6. 38 1/2 sq. yd.
7. $173.25
8. 3 1/7 sq. ft.
9. 11/14 sq. ft.

page 114

1. 2 cups flour
   3 cups sugar

   1/2 teaspoon salt

   2 2/3 cups egg whites
   3 1/3 teaspoons
      cream of tartar
   2 1/2 teaspoons vanilla

2. 1/2 cup flour
   3/4 cup sugar

   1/8 teaspoon salt

   2/3 cup egg whites
   5/6 teaspoon
      cream of tartar
   5/8 teaspoon vanilla

3. 5 cups flour
   7 1/2 cups sugar

   1 1/4 teaspoons salt

   6 2/3 cups egg whites
   8 1/3 teaspoons
      cream of tartar
   6 1/4 teaspoons vanilla

pages 116-117

1. $0.75
2. $4.74
3. $7.39
4. $0.92
5. $14.10
6. $5.57
7. $75.15
8. $4.10
9. $1.58
10. $41.17
11. $2.06
12. $0.90
13. $15.87
14. $13.19

page 119

1. $80
2. $67.50
3. $11
4. $67
5. $37.50
6. $172.50
7. $79.80
8. $57

page 120

1. $18
2. $22.50
3. $105
4. $7.20
5. $14.40

page 121

1. $162
2. $275
3. $31.50
4. $525
5. $42

page 123

1. end of first quarter — $812
   end of second quarter — $824.18
   end of third quarter — $836.54
   end of the year — **$849.09**
2. end of first quarter — $1224
   end of second quarter — $1248.48
   end of third quarter — $1273.45
   end of the year — **$1298.92**

page 125

1. $0.34 1/2
2. $0.33
3. $0.29 1/2
4. Sam's Store
5. $0.14 2/7
6. $0.14 5/6
7. $0.12 1/4
8. Sam's Store
9. $0.34 1/2
10. $0.29 1/2
11. $0.33 1/3
12. Gert's Groceries

page 127

1. 38%
2. 50%
3. 42 6/7 %
4. 27 3/11 %
5. 29 17/27 %
6. 46 2/3 %
7. 50%
8. 22 2/9 %
9. 20%
10. 24%
11. 17 1/7 %

pages 128-129

1. $19.50
2. $58.50
3. $16.65
4. $33.30
5. $150.30
6. $162.32

7.

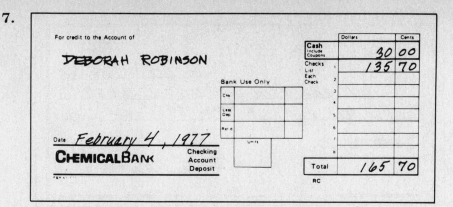

For credit to the Account of

DEBORAH ROBINSON

Bank Use Only

Chk		
Less Dep		
Ret'd		

Units

Date **February 4, 1977**

CHEMICALBANK Checking Account Deposit

	Dollars	Cents
Cash Include Coupons	30	00
Checks List Each Check 1	135	70
2		
3		
4		
5		
6		
7		
8		
Total	165	70

RC

8.

No. **004**

Feb. 6 19 **77** 1-12/210

PAY TO THE ORDER OF *The Friendly Furniture Store* $ **162.32**

One Hundred Sixty-two and thirty-two/100 _____ DOLLARS

CHEMICALBANK

FOR _____

Deborah Robinson

CHECK NO.	DATE	CHECK ISSUED TO	AMOUNT OF CHECK		DATE OF DEP.	AMOUNT OF DEPOSIT		BALANCE	
								285	00
001	1/21	Telephone Company	35	70				249	30
002	1/29	Utility Company	15	36				233	94
003	2/1	Rental Agent	145	00				88	94
004	2/6	The Friendly Furniture Store	162	32	2/4	165	70	92	32

page 131

1. $200.80	**2.** $425.10	**3.** $998.20
4. $1048.80	**5.** $1371.76	**6.** $2047.50
7. $2510.20	**8.** $3503.60	**9.** $3865.50
10. $5042.20		

pages 132-133

1. $1708.05	**2.** $559.70	**3.** $395.65
4. $135.10	**5.** $6851.50	

6.

No. 021

April 15 19 77 $\frac{1\text{-}12}{210}$

PAY TO THE ORDER OF _Internal Revenue Service_ $ 106.95

One Hundred Six and ninety-five/100 ————— DOLLARS

CHEMICALBANK

FOR _____   _Peter Munro_

7. $106.95

8.

COLWER PRODUCTS
333 EAST 49 TH STREET
N.Y., N.Y. 10017   13-5009340

Type or print EMPLOYER'S Federal identifying number, name, address, and ZIP code above.

Wage and Tax Statement 1981

Employer's State identifying number 21-49521	Copy A For Internal Revenue Service Center

FEDERAL INCOME TAX INFORMATION		SOCIAL SECURITY INFORMATION		STATE OR LOCAL INCOME TAX INFORMATION			
1 Federal income tax withheld	2 Wages, tips, and other compensation	3 FICA employee tax withheld	4 Total FICA wages	6 Tax withheld	7 Wages paid	8 State or locality
$1708.05	$9650.00	$559.70	$9650.00	$395.65	$9650.00	NEW YORK

EMPLOYEE'S social security number ▶ 123-45-6789	5 Uncollected employee FICA tax on tips NONE	9 Tax withheld $135.10	10 Wages paid $9650.00	11 State or locality NEW YORK

PETER MUNRO
5 EAST THIRD STREET
N.Y., N.Y. 10003

OTHER INFORMATION — STATUS

Was employee covered by a qualified pension plan etc.? Yes ☐ No ☐	Contribution to individual employee retirement account	Cost of group term life insurance included in box 2	Excludable sick pay included in box 2	STATUS 1. Single 2. Married

If this is a corrected form, put an "X" to the right of the number in the upper left corner.

Type or print EMPLOYEE'S name, address, and ZIP code above.

For instructions see Form W-3 and back of Copy D.

Form **W-2**   Department of the Treasury—Internal Revenue Service

pages 134-135

1. $2205	**2.** $8295	**3.** $1659
4. $138.25	**5.** $2571.45	**6.** $49.45
7. $414.75	**8.** $497.70	**9.** $829.50
10. $207.38	**11.** $995.40	**12.** $829.50
13. $497.70		

pages 136-138

1. (a) $171.72
(b) $195.90
(c) $24.18

2. (a) $418.70
(b) $479.25
(c) $60.55

3. (a) $481.95
(b) $582.72
(c) $100.77

4. (a) $307.35
(b) $360.39
(c) $53.04

5. (a) $120.23
(b) $168.32
(c) $48.09